Topology of Surfaces, Knots, and Manifolds: A First Undergraduate Course

1ST EDITION

Topology of Surfaces, Knots, and Manifolds: A First Undergraduate Course

STEPHAN C. CARLSON
Rose-Hulman Institute of Technology

JOHN WILEY & SONS, INC.
New York / Chichester / Weinheim / Brisbane / Singapore / Toronto
http://www.wiley.com/college

Acquisitions Editor   *Mary Johenk*
Marketing Manager   *Julie Lindstrom*
Senior Production Editor   *Michael Farley*
Senior Designer   *Kevin Murphy*
Production Management Services   *Publication Services*

This book was set in *Times Roman* by *Publication Services* and printed and bound by *Courier Westford*. The cover was printed by *Lehigh Press*.

The book is printed on acid-free paper.

Library of Congress Cataloging-in-Publication Data:
Carlson, Stephan C.
    Topology of surfaces, knots, and manifolds: a first undergraduate course / Stephan C. Carlson.—1st ed.
        p.   cm.
    Includes bibliographic references.
    ISBN 0-471-35544-5 (cloth: alk. paper)
    1. Topology. I. Title

QA611 .C328 2000
514—dc21

                                                                                    00-063294

Printed in the United States of America
10 9 8 7 6 5 4 3 2 1

To my son, Alex.

Preface

This book's six chapters consist of an appealing set of ideas and problems involving curves, surfaces, and knots. These notions, which reside at the core of topology as envisioned by the subject's pioneers in the late nineteenth century, have been explored extensively by mathematicians during the twentieth century. Now in the twenty-first century, mathematicians and scientists are collaborating to apply some of these very ideas in areas such as physics, chemistry, and biomedical research. These ideas and problems represent the intriguing "rubber sheet geometry" description of topology, which a student might have encountered in secondary school or earlier and hoped to learn more about some day. They have also motivated much of the development of general topology which, in the traditional mathematics major curriculum, has become a prerequisite for studying combinatorial properties of manifolds, a study that usually takes place at a quite rigorous level in a course on algebraic topology or differential geometry. It has, therefore, been the case that advanced mathematics majors—usually those that are headed for graduate study—form the primary group of students who have had the opportunity to take a course which formally develops the wonderful subject matter of "rubber sheet geometry" topology.

This book is intended to serve as a textbook for a one-semester topology course offered to undergraduate students who have not previously studied the subject. Its aim is to introduce some basic ideas and problems concerning manifolds, especially one- and two-dimensional manifolds, via an intuition-based and example-driven approach. A course based on this book can be offered to any undergraduate student with a calculus background, although a course or two beyond calculus, such as linear algebra or discrete mathematics, might be helpful. Such a course could, therefore, be an early introduction to topology offered to mathematics majors before they enroll in traditional upper division courses, and it would be very appropriate for mathematics majors preparing for careers in secondary school mathematics teaching. Also, essentially any student in a technical major—not just a mathematics major—should be able to succeed in an elective course based on this book.

The approach of this book, which makes the topology of manifolds accessible to a broad group of undergraduate students, requires some compromising of rigor for conceptual understanding. But it neither ignores the solid mathematics underlying the topics nor gives up the expectation that students should do some real mathematical work and work hard at mathematical discovery. Fortunately, only the topology of Euclidean space is required to investigate manifolds in this book (in fact, four dimensions suffice throughout), and the point-set topological ideas can be handled successfully in a relatively intuitive fashion.

This book has grown out of course notes that I have used successfully in introductory topology courses during the past two decades. Of course, I have been inspired and influenced by books written by other authors who have chosen to present the topics of combinatorial topology and manifolds, often intuitively, to an undergraduate audience. Among these books are some whose main focus is a specialized area (e.g., knot theory) or whose aim is to reach more advanced topics (for example, geometries on 3-manifolds, group theory, or homology). Others emphasize a "discovery" approach using projects and well-designed exercises to lead the reader to general results with less emphasis on a unifying theory or associated proof techniques. In this book, I have pursued a course which lies between those mentioned above. The general unifying theme here is the topology of manifolds, and the main presented results concerning surfaces and knots are well within the grasp of the intended audience. A blend of examples and exercises leads the reader to anticipate general definitions and theorems whose proofs are provided when appropriate, or discussed in an honest and open way when the technical details lie beyond the book's intended scope.

I wish to note especially the books that have had the greatest influence on my treatment of the topics in this textbook. These include *Surface Topology* by P. A. Firby and C. F. Gardiner (Ellis Horwood Limited; 1982, 1991), *A Combinatorial Introduction to Topology* by Michael Henle (W. H. Freeman, 1979; Dover, 1994), *Knot Theory* by Charles Livingston (Mathematical Association of America, 1993), *The Shape of Space* by Jeffrey R. Weeks (Marcel Dekker, 1985), and *Introduction to Graph Theory* by Robin J. Wilson (Academic Press, 1972). These books, along with a number of other resources, are listed in the "References and Suggestions for Further Study" section. A reference in the present text to any of these sources is indicated by the alphabetically first author's last name set in bold face within brackets (e.g., [**Firby**] refers the reader to *Surface Topology* by P. A. Firby and C. F. Gardiner).

The layout and format of this book have been designed to maximize the success of students in mastering the topics covered. Some of the special aspects of the design of this book include the following.

- A friendly writing style enhances a clear exposition of the topics covered.

- A proper and correct presentation of the mathematics, including clearly labeled statements of definitions, theorems, and main results, is offered in the context of an inviting discussion with the reader.

- Extensive use of illustrations supports building of intuition and complements understanding of examples.

- Frequent historical references concerning the development of the subject matter reinforce the fact that mathematics is a human endeavor and emphasize the connections between topology and other areas of mathematics.

- Chapter Summary sections review and highlight topics covered in each chapter and offer a transitional look forward to the next chapter.

- "Point of encounter" exercises included in each chapter provide readers immediate opportunities to reinforce understanding of concepts or to develop interesting examples.

- Sets of Supplementary Exercises at the end of each chapter offer further investigation of topics covered or introduce related ideas.

- Exercises that require "hands-on" work with models constructed by the reader or computer-aided visualization are included.
- A "References and Suggestions for Further Study" section offers readers some direction for continued learning beyond this book.

I would like to thank my mathematical colleagues at Ball State University, Indiana University, and the University of North Dakota—who have adopted preliminary versions of this textbook for use in their classes—as well as the mathematics editorial staff at John Wiley & Sons, Inc., for invaluable comments and advice during the preparation of the manuscript. I also wish to express my gratitude for the many suggestions and constructive criticism provided by the following reviewers: Nathanael B. Berglund, Rose-Hulman Institute of Technology; Carlos C. Borges, University of California, Davis; John W. Emert, Ball State University; Chuck Livingston, Indiana University; Jerry Metzger, University of North Dakota; Roger B. Nelson, Ball State University; Jack Porter, University of Kansas; Nicholas A. Robarge, Rose-Hulman Institute of Technology; Russell J. Rowlett, University of North Carolina, Chapel Hill; Gary L. Walls, University of Southern Mississippi; and Stephen J. Young, Rose-Hulman Institute of Technology.

Finally, I would like to extend a huge expression of thanks to my former and current students who have always been my best critics and provided the inspiration that led to the writing of this book.

Stephan C. Carlson
Department of Mathematics
Rose-Hulman Institute of Technology
Terre Haute, IN 47803

Contents

Topology of Surfaces, Knots, and Manifolds: A First Undergraduate Course

Chapter 1

Introduction and Intuitive Ideas

1.1 WHAT IS TOPOLOGY?

Many problems in mathematics and science involve ideas that are more qualitative than quantitative. Molecular biologists are interested in a qualitative description of how DNA (viewed as two long intertwined curves) is wound around enzymes and how this structure is related to the dynamics of the enzyme actions. A physicist or engineer might want to know the shape of the solution curves in the phase portrait of an oscillation problem when an analytic description of the solutions is impossible to obtain. These are just two examples of the many problems whose solutions have been aided in recent decades by methods of the branch of mathematics called *topology*.

The ideas exhibited by the two previous examples suggest that topology is geometrical, in that an understanding of certain shapes seems to be at the heart of these problems. But unlike geometry, in which shapes are considered "equivalent" if they can be transformed into each other in a rigid fashion (leading to the ideas of similarity and congruence), topology allows a more relaxed idea of equivalence of shapes.

Here is a simple, useful, and descriptive definition of topology.

Definition

Topology is the branch of mathematics dealing with properties of objects that are unaffected by continuous deformation.

This definition is sufficiently vague to be useful in a general setting! The vagueness should be clear to you if you re-read the words a few times. You are then forced to ask what is meant by "object" and by "continuous deformation." These are questions we of course must answer, and we will do so several times—with some increasing degree of rigor as we go along. Once these ideas are made clear, we expect the "properties" mentioned in the definition to be what we will be learning about throughout this book.

By *object* we will mean a set of points in space. For a transformation of one object into another to be a *continuous deformation,* we allow it to involve such

1

motions as translation, bending, stretching, shrinking, and twisting, but we do *not* allow any tearing apart of the object or gluing together of portions of the object. Looking at some simple examples now will help motivate these ideas, but first we introduce a preliminary topological analogue of the geometrical notions of similarity and congruence.

Definition

If a first object can be continuously deformed so as to coincide with a second object, then the two objects are said to be *isotopic* (or *isotopic to each other*).

It should be noted that this definition is somewhat preliminary, in that we will be discovering that some deeper related conditions must be assumed. The understanding of "isotopic" we will eventually adopt is often called "ambiently isotopic" in more formal and rigorous treatments of topology. But we shall be using the simpler phrase "isotopic" throughout this book, tacitly including the "ambient" conditions that are yet to be discussed.

EXAMPLE 1.1

In Figure 1.1, the closed planar curves in the shapes of a circle, a square, and a triangle, are pairwise isotopic objects. It is not difficult to see this if we imagine the circle as a rubber band that can be continuously deformed in the plane into either the square or the triangle. Of course, this rubber band model only approximates the actual idealized mathematical objects that, being closed curves, have zero thickness. In Figure 1.2, a sequence of sketches demonstrates the steps of a continuous deformation in the plane of a more complicated tangle into a circle.

EXAMPLE 1.2

The surface of a cube is isotopic to a solid ball's surface, which is usually called a *sphere*. We can imagine these surfaces, shown in Figure 1.3, as being made of rubber sheets that can be deformed one into the other. Again we must realize that these idealized mathematical surfaces have zero thickness, unlike the rubber sheets we are using as a model.

Figure 1.1 Three pairwise isotopic curves.

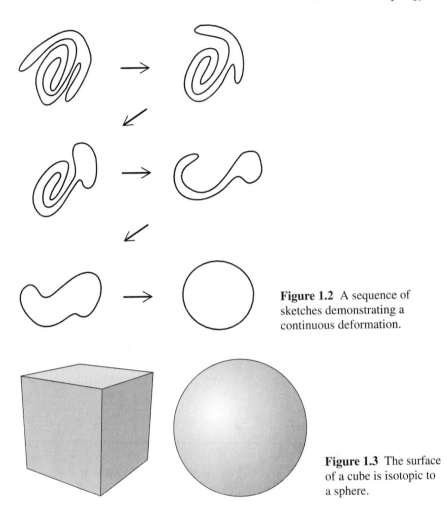

Figure 1.2 A sequence of sketches demonstrating a continuous deformation.

Figure 1.3 The surface of a cube is isotopic to a sphere.

EXAMPLE 1.3

A sphere is *not* isotopic to a *torus* (the surface of a solid donut). This seems reasonable when we look at those surfaces illustrated in Figure 1.4: The torus has a hole in it (the "donut hole") while the sphere has no such hole. But an argument that two objects are not isotopic requires a more rigorous approach. Note that the sphere has the property that any closed loop on it can be continuously deformed, while kept on the surface, so that it has as small a diameter as we desire and approaches a point on the sphere. (We say that such a loop can be "shrunk to a point" on the sphere.) And if another object is isotopic to the sphere, it will have this property too. It is easy to see that one can find closed loops on a torus that cannot be shrunk down to an arbitrarily small diameter, so a torus is not isotopic to a sphere.

EXAMPLE 1.4

Various surfaces that are isotopic to a torus are shown in Figure 1.5. The surface of a simple picture frame is interesting in that the "torus" surface is divided up into a number

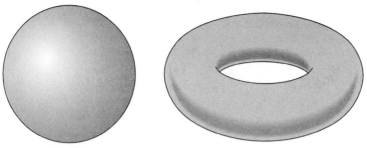

Figure 1.4 A sphere is not isotopic to a torus.

of four-edged "faces." The surface of an ordinary coffee mug illustrates the adage (popular in mathematical circles) that asserts that "a topologist (a mathematician who studies topology) can't tell a coffee mug from a donut!"

EXAMPLE 1.5

By *cylinder* we will mean an object isotopic to a "tin can" surface with the top and bottom cut off. One technical mathematical requirement we will make in this book for a cylinder is that the circular boundaries of its top and bottom are considered to be *included* in the cylinder. A method for constructing a paper cylinder will be useful later in a general context. Start with a rectangular strip of paper and bend it into a cylindrical shaped roll, finally gluing or taping two opposite ends together. Here we are allowing gluing to make the cylinder from the rectangle. Recall that gluing is not allowed in a continuous deformation, but we are *not* asserting that the rectangle and cylinder are isotopic—we are just constructing the cylinder. In fact, the rectangle and the cylinder are not isotopic, and the reader should be able to give an argument to this effect.

Figure 1.5 Three isotopic surfaces.

• **Exercise 1.1** The cylinder, which we usually view as an object in 3-dimensional space, is isotopic to a subset of the plane—namely a plane annulus, the region between or on two concentric circles. Carefully draw a sequence of sketches illustrating the steps of a continuous deformation of a cylinder into an annulus.

EXAMPLE 1.6

A variation on the method used to construct a paper cylinder in the previous example can be used to construct one of the most famous objects in mathematics—the *Möbius band*. Again, start with a rectangular strip of paper, this time somewhat longer between the left and right ends than between the top and bottom edges. Bend the paper as before, but before gluing the left and right ends together, give the strip a half-twist. The resulting object—the Möbius band—is not isotopic to the cylinder. A Möbius band is illustrated, along with a cylinder, in Figure 1.6. For both the cylinder and the Möbius band, the boundary (which is part of both objects) is formed by the top and bottom edges of the paper strip we started with. A careful inspection will reveal that these edges combine on the Möbius band to form a boundary set consisting of one closed curve in space, while the boundary set of the cylinder includes two disjoint closed curves. If two surfaces, including their boundaries, are isotopic, then their boundary sets would also be isotopic. So the cylinder and the Möbius band are not isotopic. We will investigate properties of the Möbius band later; it is one of the best-known examples of a "one-sided surface."

• **Exercise 1.2** It is intuitively clear that two linked tori can not be unlinked in 3-space. However, if the linked tori are joined by a tube, as shown on the left in Figure 1.7, the resulting object *can* be continuously deformed *in 3-space* into the object shown on the right in the figure in which the tori appear unlinked! Demonstrate such deformation with a sequence of sketches.

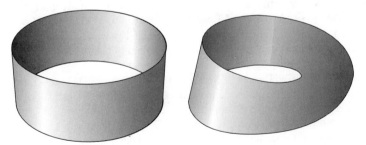

Figure 1.6 The cylinder and the Möbius band are not isotopic.

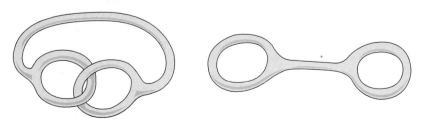

Figure 1.7 Two surprisingly isotopic objects.

The sets of points considered in all of these examples are closely related to some of the types of topological objects that we will be studying throughout this book. These objects are called *manifolds* (terminology that will be defined officially later) and include the special important cases of knots and surfaces. The study of the topological properties of knots and surfaces is at the heart of the genesis of the subject of topology and continues even today as an important research area with many applications in the sciences. Moreover, the properties of these special manifolds have motivated much of the subject matter of the more general areas of point-set topology, geometric topology, and algebraic topology and have influenced the development of differential geometry, the algebraic study of groups, and many other areas of mathematics.

Historically, it is often said that topology officially became a field of mathematical study with the publication in 1895 of *Analysis Situs* by the French mathematician Henri Poincaré. ("Analysis Situs" is a Latin phrase meaning "analysis of position.") Before moving on to the next somewhat more technical section, the reader may wish to review some of the interesting historical timeline (see Table 1.1) leading up to Poincaré's important publication. Many of the problems mentioned here will be discussed later in this book.

Table 1.1   Historical Time Line

1736: Swiss-born mathematician Leonard Euler (1707–1783) publishes his solution to the problem of the seven bridges of Königsberg. He classifies his solution as an example of "geometria situs" (the geometry of position), terminology introduced earlier by Gottfried Leibniz (1646–1716), who, along with Isaac Newton, was one of the "coinventors" of calculus.

1750: Euler uses the formula $V - E + F = 2$, which would later be called the "Eulerian polyhedral formula." This formula, which we shall study in a more general form, was known to René Descartes (1596–1650) as early as 1640.

1847: German astronomer J. B. Listing (1808–1882) publishes *Vorstudien zur Topologie* (*Preliminaries to Topology*). This is generally regarded as the first usage of the word "topology."

1851: German mathematician Bernhard Riemann (1826–1866) submits his doctoral thesis on complex number theory and what today are called "Riemann surfaces," thus introducing topological considerations into analysis. Riemann's thesis director was C. F. Gauss (1777–1855). The thesis also contained ideas on the definition of the integral and the Cauchy-Riemann equations.

1852: English graduate student Francis Guthrie (1831–1899) poses the famous four-color problem for planar maps. This problem was "solved" in 1879 by A. B. Kempe (1849–1922), but an error was found in his proof by P. J. Heawood (1861–1955). Heawood went on to pose and solve the corresponding five-color problem and to consider coloring of maps on various surfaces. The four-color problem was ultimately solved in 1977 by American mathematicians K. Appel and W. Haken of the University of Illinois.

(*Continued*)

Table 1.1   *(Continued)*

1858:	German astronomer and geometer A. F. Möbius (1790–1868) submits a memoir on "one-sided" surfaces to the Paris Academy. His work, done independently, nevertheless contained ideas similar to some of those in Listing's *Vorstudien*. We have already seen an example of a one-sided surface which is actually named after Möbius. It is alleged that Möbius posed the four-color problem in 1840, twelve years earlier than Guthrie.
1872:	German geometer Felix Klein (1849–1925) presents his inaugural address as professor at Erlangen in which he places topology among geometries as "the theory of invariants of continuous point transformations."
1882:	Klein gives an exposition on Riemann's work and presents an example of a one-sided "closed" surface, now known as the *Klein bottle*. Klein, an inspiring teacher who was loved and admired by his students, is known as a pioneer mathematics educator.
1882:	French mathematician Camille Jordan (1838–1922) publishes *Cours d'Analyse* containing his rigorous proof of an intuitively obvious topological result about curves in the plane. His recognition of the need for such rigor set a precedent for much of modern mathematics. Today that result is known as the "Jordan curve theorem."
1895:	French mathematician Henri Poincaré (1854–1925), who would be the dominant figure in the world of mathematics at the beginning of the twentieth century, publishes *Analysis Situs*. Poincaré, who explored many special surfaces as potential domains for solutions to differential equations, took on the problem of classifying surfaces and, in doing so, became "the father of modern topology." Proving or disproving his famous conjecture on the nonexistence of "fake" 3-spheres—the "Poincaré conjecture"—remains a major challenge to today's mathematicians.

1.2   GETTING MORE FORMAL

In this section we will introduce some basic ideas that will be needed throughout the remainder of our investigation. In one sense, we are lucky: Our study of properties of curves, surfaces, and other manifolds will require only the topological ideas of Euclidean spaces. Thus many of the definitions may be presented in quite simple ways. On the other hand, we should remember that our definitions generalize in a natural way to a more abstract setting, and it is in this setting that mathematicians usually study topological notions.

One-dimensional Euclidean space (or simply 1-*space*) is just the set of real numbers. The Euclidean distance between two numbers x_1 and x_2, thought of as points on a straight line, is $|x_1 - x_2|$. *Two-dimensional Euclidean space* (2-*space*) is the set of all ordered pairs (x, y) of real numbers x and y. We identify the elements (ordered pairs) of 2-space with the points in the familiar *x-y* plane, and 1-space can be thought of as sitting inside 2-space as the *x*-axis when we identify a real number x with the ordered pair $(x, 0)$. The usual *Euclidean distance function* for 2-space defines the distance between the points (x_1, y_1) and (x_2, y_2) to be $\sqrt{(x_1 - x_2)^2 + (y_1 - y_2)^2}$. Similarly, *three-dimensional Euclidean space* (*3-space*) is the set of all ordered triples (x, y, z) of real numbers which we identify with the points in familiar x-y-z space in which we live. The Euclidean distance function in 3-space declares the distance between the points (x_1, y_1, z_1) and (x_2, y_2, z_2) to be $\sqrt{(x_1 - x_2)^2 + (y_1 - y_2)^2 + (z_1 - z_2)^2}$. We can think of 2-space as a specially situated subset of 3-space if we identify the ordered pair (x, y) and the ordered triple $(x, y, 0)$. In this way, 2-space sits inside 3-space as the x-y plane.

An important (although somewhat trivial sounding) observation is that the distance between two points of 2-space is the same whether measured using the 2-space distance function or the 3-space distance function once the appropriate identification is made. Similarly, the distance between two points of 1-space is the same whether measured using the 1-space, 2-space, or 3-space distance function. For now, we will use the word *space* to refer to Euclidean 3-space. The objects whose properties we will study in this book will be certain types of sets of points in space. An advantage of having Euclidean space coordinates is that we can now use equations or inequalities to describe particular objects of interest.

EXAMPLE 1.7

Intervals are important subsets of the real numbers, that is, of 1-space. For real numbers a and b, with $a < b$, we will use the following notation.

$$(a, b) = \{x : a < x < b\}$$
$$[a, b] = \{x : a \le x \le b\}$$
$$[a, b) = \{x : a \le x < b\}$$
$$(a, b] = \{x : a < x \le b\}$$

We call (a, b) an "open interval," $[a, b]$ a "closed interval," and $[a, b)$ and $(a, b]$ "half-open intervals." We will also use the usual notation for "infinite length" intervals.

$$(a, \infty) = \{x : x > a\}, [a, \infty) = \{x : x \ge a\}$$
$$(-\infty, b) = \{x : x < b\}, (-\infty, b] = \{x : x \le b\}$$
$$(-\infty, \infty) = \{x : x \text{ is any real number}\}$$

EXAMPLE 1.8

The *unit circle* is the set of points in the plane whose distance from the origin is exactly one unit. So this is the set of points (x, y) satisfying the equation $x^2 + y^2 = 1$. The *open unit disk* is the set of points strictly inside the unit circle and is, therefore, described by the strict inequality $x^2 + y^2 < 1$. The *closed unit disk* contains all points on or inside the unit circle; its points (x, y) satisfy the nonstrict inequality $x^2 + y^2 \le 1$. Of course, other circles and disks with various centers and radii can be described using similar equations and inequalities.

EXAMPLE 1.9

The *unit sphere* in space is the set of points in space at exactly a distance of one unit from the origin. Hence, its points (x, y, z) satisfy the equation $x^2 + y^2 + z^2 = 1$. The open (or closed) unit *ball* would be described by the corresponding strict (or nonstrict) inequality involving x, y, and z.

EXAMPLE 1.10

A particular cylinder in space of radius one unit and height two units can be described as the set of all points (x, y, z) satisfying both the equation $x^2 + y^2 = 1$ and the inequality $-1 \leq z \leq 1$.

It is sometimes useful to introduce parameters in terms of which we may express the space variables x, y, and z.

EXAMPLE 1.11

Let t be a parameter on which x, y, and z depend according to the parametric equations

$$x = \cos t, y = \sin t, z = \cos 2t \ (0 \leq t \leq 2\pi)$$

The resulting set of points (x, y, z) in space is a curve isotopic to the unit circle and is shown in Figure 1.8.

EXAMPLE 1.12

The introduction of two parameters allows a parametric description of a surface. Let s and t be two parameters, and let x, y, and z be given by the equations

$$x = 2\sin s \cos t, y = 2\sin s \sin t, z = 2\cos s (0 \leq s \leq \pi/2, 0 \leq t \leq 2\pi)$$

The resulting graph in space is a closed upper hemisphere of radius 2, shown in Figure 1.9. (The careful reader will recognize the parameters s and t to be the spherical coordinates φ and θ, respectively.) This object is isotopic to the closed unit disk in the plane.

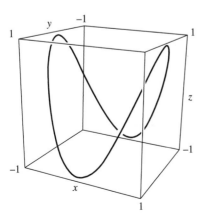

Figure 1.8 A curve in space that is isotopic to the circle.

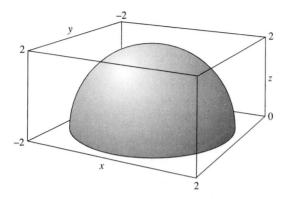

Figure 1.9 A surface isotopic to the closed unit disk.

EXAMPLE 1.13

Let us make use of parametric equations to describe one of the surfaces already encountered—the torus. One specific representation could take the torus to be a surface of revolution obtained by revolving a circle in the right half of the x-z plane about the z-axis. In each half plane containing the z-axis and making an angle of θ radians with the positive x-axis, the intersection with this torus would be an identical circle, as shown in Figure 1.10.

Referring to this figure, we see that each such plane contains perpendicular unit vectors **k** and **u** $= \cos\theta\,\mathbf{i} + \sin\theta\,\mathbf{j}$. The vector **r** which gives the position of the general point P on the torus in terms of θ and the angle γ is

$$\mathbf{r} = a\mathbf{u} + \mathbf{v}$$
$$= a\mathbf{u} + b(\cos\gamma\,\mathbf{u} + \sin\gamma\,\mathbf{k})$$
$$= (a\cos\theta + b\cos\theta\cos\gamma)\,\mathbf{i} + (a\sin\theta + b\sin\theta\cos\gamma)\,\mathbf{j} + (b\sin\gamma)\,\mathbf{k}$$

Thus, this torus is described in terms of parameters θ and γ by the equations

$$\begin{cases} x = a\cos\theta + b\cos\theta\cos\gamma \\ y = a\sin\theta + b\sin\theta\cos\gamma \quad (0 \le \theta \le 2\pi, 0 \le \gamma \le 2\pi) \\ z = b\sin\gamma \end{cases}$$

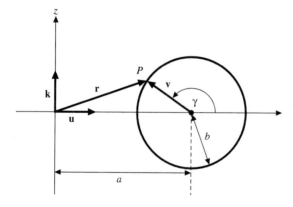

Figure 1.10 A circle generates a torus as a surface of revolution.

With the aid of a computer graphing program or computer algebra system, one can choose specific values of a and b and obtain a very pleasing graph of a torus.

• **Exercise 1.3**   Mimic the method used in Example 1.13 to find parametric equations for a cylinder of radius a centered on the z-axis and extending from $z = -b$ to $z = b$. Then use any available computer algebra system (or computer graphing system) to produce a graph of the cylinder from your parametric equations for specific values of a and b. (Hint: Start by drawing a picture of the half-plane that contains the z-axis and makes an angle θ with the positive x-axis and in that plane show the vertical line that is the intersection of the cylinder with this plane. Use θ and z as parameters.)

• **Exercise 1.4**   Appropriately modify the work you did in Exercise 1.3 to obtain parametric equations for a Möbius band having as its "center circle" the circle $x^2 + y^2 = a^2$ in the x-y plane and having "band width" $2b$. Again, use a computer algebra system or graphing system to produce a graph of a Möbius band from your parametric equations.

1.3   FUNCTIONS AND HOMEOMORPHISMS

Armed with the structure of space—and especially its Euclidean distance—we can now introduce a precise notion of topological equivalence of objects. Let X and Y be two objects (i.e., sets of points in space). Recall that a function f from the *domain* X to the *codomain* Y is a rule that assigns to each point x of X one and only one point y, also denoted by $f(x)$, of Y. We say that "x gets mapped to y by f." Such a function is a *bijection* or *one-to-one correspondence* if it is one-to-one (meaning that $f(x_1) \neq f(x_2)$ whenever $x_1 \neq x_2$) and is also onto (for each point y of Y, there is a point x of X such that $y = f(x)$). (This latter condition makes Y the *range* of f.) Note that a bijection f from X to Y has an inverse function f^{-1} with domain Y and range X defined by $x = f^{-1}(y)$ if and only if $y = f(x)$.

EXAMPLE 1.14

Consider the function rule $f(x) = x^2$, familiar to every calculus student, where x denotes a real number. If we use $X = (-\infty, \infty)$ as the domain of f and $Y = (-\infty, \infty)$ as the codomain of f, then f is not one-to-one (since, for example, 2 and –2 have the same square, namely 4) and f is not onto (since, for example, the number –1, which does belong to Y, is not the square of any real number). However, if we specify $Y = [0, \infty)$ as the codomain associated with the domain $X = (-\infty, \infty)$, then f becomes an onto function. And if we further specify both $X = [0, \infty)$ as domain and $Y = [0, \infty)$ as codomain, then f is both one-to-one and onto or, in other words, is a bijection or a one-to-one correspondence. Note that, in this last case, the inverse function f^{-1} is the square root function with rule $f^{-1}(x) = \sqrt{x}$, domain $[0, \infty)$, and range $[0, \infty)$.

• **Exercise 1.5**   Determine whether each of the following choices of domain X and codomain Y for the function rule $f(x) = x^2$ makes sense and, for each that does make sense, decide whether the function f is one-to-one, whether f is onto, and whether the function f is a bijection.

(a) $X = [1, \infty)$ and $Y = (-1, \infty)$

(b) $X = [-2, 2]$ and $Y = [0, 1]$

(c) $X = [-2, 0]$ and $Y = [0, 4]$

(d) $X = [0, 1)$ and $Y = (0, 1]$

The notion of continuity of real-valued functions of a real variable should be familiar to readers who have studied calculus, at least on an intuitive level. For example, suppose for the function $f(x) = x^2$ we have specified some meaningful choices of sets of real numbers X and Y as domain and codomain, respectively. We may argue that f is continuous by noting that for any real number p in X and for any positive distance ε, there is some corresponding distance δ such that if q is in X and the distance between p and q is less than δ then the distance between p^2 and q^2 is less than ε. A little less formally, we might note that this continuity really means that numbers that are close to each other have squares that are also close; and, from an intuitive graphical point of view, we observe that if the domain of the squaring function is an interval, then the function's graph can be drawn as an unbroken piece of a parabola.

We will need to address the continuity of functions with domains and codomains that are objects consisting of sets of points in space, not just sets of real numbers. Our definition of continuous function is one that is precise yet retains the intuitive idea of close points mapping to close points.

Definition

Let X and Y be sets of points in space, and let f be a function with domain X and codomain Y. Then f is *continuous* if whenever a sequence $\{x_n\}$ of points in X converges to (i.e., gets "closer and closer to") a point x of X, then also the sequence $\{f(x_n)\}$ of points in Y converges to the point $f(x)$, where "closeness" is measured by the Euclidean distance function.

Figure 1.11, which shows the graph of $f(x) = x^2$, illustrates a sequence $\{x_n\}$ of real numbers along the x-axis converging to the number 1 and the corresponding sequence $\{f(x_n)\}$ of numbers along the y-axis converging to $f(1) = 1^2 = 1$.

EXAMPLE 1.15

Consider the subset $X = [0, 3]$ of the real line and let Y be the graph of $f(x) = x^2$ defined on X as its domain. Recall that by the graph of f we mean the set of all points (x, y) in the plane for which $y = f(x)$ and x belongs to the domain X. Thus, $Y = \{(x, y) : y = x^2 \text{ and } 0 \leq x \leq 3\}$. Let us define a function g with domain X and codomain Y by the rule $g(x) = (x, x^2)$. It is easy to check that the function g is a bijection. The continuity of g can be seen by observing that if x_n gets close to x, then also x_n^2 must get close to x^2 (by the continuity of the squaring function) and so the point $g(x_n) = (x_n, x_n^2)$ must get close to the point $g(x) = (x, x^2)$. Figure 1.12 shows a sequence $\{x_n\}$ of numbers in $[0, 3]$ converging to 1 and the corresponding sequence $\{g(x_n)\}$ of

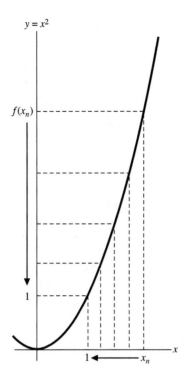

Figure 1.11 The squaring function is continuous.

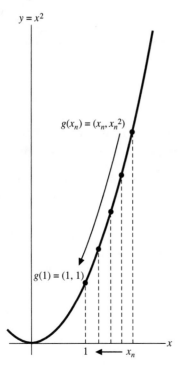

Figure 1.12 The function g is continuous.

points on Y converging to the point $g(1) = (1, 1)$. The reader should carefully note the differences between what is illustrated in Figures 1.11 and 1.12.

In this case, the inverse of the bijection g is the function g^{-1} with domain Y, the graph of the squaring function, and codomain $X = [0, 3]$, and the rule for g^{-1} is $g^{-1}(x, y) = x$. This function, which "projects" a point (x, y) onto its first coordinate x, is called the *first coordinate projection function*. (There is, of course, a *second coordinate projection function* also.) The continuity of the function g^{-1} is clear; if points (x_n, y_n) in the plane approach a certain point (x, y), then the first coordinates x_n approach that point's first coordinate x. Therefore, the function g of this example is a continuous bijection for which the inverse function is also continuous.

The domain X and the codomain Y of the function g of Example 1.15 appear to be isotopic as objects in space. X is a closed interval of the real line, and Y is a "closed" unbroken portion of the parabola $y = x^2$. In fact, we can imagine a continuous deformation of Y into X consisting of a gradual lowering of each point of Y along a vertical line. This continuous deformation in fact induces the one-to-one correspondence given between the points of X and Y by the bijection g. With this in mind, we now define a notion of topological equivalence in terms of continuous functions.

Definition

Let X and Y be two sets of points in space. We will say that X and Y are *homeomorphic* if there is a continuous bijection from X to Y that has a continuous inverse.
  We call such a bijection a *homeomorphism* from X to Y.
  We will say that two objects are "topologically equivalent" if they are homeomorphic.

Thus, the function g of Example 1.15 is a homeomorphism, and its domain $[0, 3]$ and range $\{(x, y) : y = x^2 \text{ and } 0 \le x \le 3\}$ are homeomorphic objects in space. In fact, the arguments of that example may be translated to confirm that the graph of any continuous real-valued function of a real variable is homeomorphic to its domain.

It is a seemingly simple, but important, observation that if X is homeomorphic to Y, then also Y is homeomorphic to X. Indeed, if h is a homeomorphism from X onto Y, then h^{-1} is a homeomorphism from Y onto X.

• **Exercise 1.6**   Determine the possible pairs of numbers m and b which make the linear function with rule $h(x) = mx + b$ a homeomorphism from domain $[-3, 5]$ to range $[2, 8]$, and for each possibility find the rule for the inverse of h. Note that h and h^{-1} will be continuous since linear functions are continuous.

• **Exercise 1.7**   Show that $X = \{(x, y) : 0 < x^2 + y^2 < 1\}$ (a "punctured open disk") is homeomorphic to $Y = \{(x, y) : 1 < x^2 + y^2 < 4\}$ (an "open annulus") by explicitly defining a homeomorphism with domain X and range Y. Give a brief argument showing that the function you define really is a homeomorphism.

We have already observed that in Example 1.15 there appeared to be a relationship between topological equivalence of two objects—being homeomorphic—and the notion of them being isotopic. If one can produce a continuous deformation in space of a first object into a second, then it can be used to suggestively produce a homeomorphism: Assign to a point of the first object the point of the second object to which it moves under the deformation. We will let this intuitive argument suffice as a proof of the following result.

Theorem

If two objects are isotopic, then they are homeomorphic.

In the remainder of this section, "isotopic" and "homeomorphic" and their relationship will be clarified to some greater extent. We will turn to some examples for motivation.

EXAMPLE 1.16

The half-open interval [0, 1) and the unit circle with equation $x^2 + y^2 = 1$ certainly do not appear to be topologically equivalent. The sequence of motions of the interval shown in Figure 1.13 seems to come close to being a continuous deformation onto the circle, but a final "gluing" of the ends of the interval is problematic. (Note that in this gluing no points ever coincide, but opposite ends of the interval just come together and butt up against each other to form an unbroken circle.) More formally, this figure suggests how to define a particular "almost homeomorphism" h with domain the half-open interval and range the unit circle given by the rule $h(x) = (\cos 2\pi x, \sin 2\pi x)$. Certainly h is a continuous bijection, but h is not a homeomorphism since h^{-1} is not continuous. To see this, think about picking points on the circle that are in the fourth quadrant and arbitrarily close to the point (1, 0). The images of such points under h^{-1} are real numbers very close to 1, but the image of the point (1, 0) under h^{-1} is the number 0, which is sufficiently far from 1 to not be considered "close" to 1. Now the question remains of how to come up with a convincing argument that the interval and the circle are not homeomorphic. Previously it was argued only that a *particular* function is not a homeomorphism. This does not say that there cannot be some other function from the interval to the circle that *is* a homeomorphism. One convincing argument goes like this: Removing any interior point from the interval divides it into two component parts, but there is no such point of the circle for which removal will divide it into two parts. Such points are called *cutpoints,* and a

Figure 1.13 This bending does not produce topological equivalence.

homeomorphism from one object to another will map a cutpoint to a cutpoint. So the interval and the circle are not homeomorphic.

EXAMPLE 1.17

We can construct a homeomorphism from a sphere to the surface of a cube by an intuitive visual method. Begin with a sphere, situated completely inside of the cube surface. (Refer to Figure 1.14.) Fix a reference point P_0 inside of, but not on, the sphere. Given a point P on the sphere, draw a ray beginning at P_0 and passing through the sphere at P. Then define $h(P)$ to be the unique point at which the ray passes through the cube surface. The reader should attempt to explain why the function h satisfies all the properties of a homeomorphism. Again, this homeomorphism is suggested by a continuous deformation of the sphere into the cube surface consisting of some special type of inflation of the sphere.

EXAMPLE 1.18

Let a first object X consist of two component parts, the unit sphere along with the origin $O = (0, 0, 0)$ inside the sphere. Similarly let a second object Y consist of two component parts, the unit sphere along with the point $Q = (0, 0, 2)$ outside of the sphere. (Although X and Y share all the points of the sphere, we will still think of them as two different objects.) It is simple to define a homeomorphism from X to Y, by taking

$$h(p) = \begin{cases} p & \text{if } p \text{ is on the unit sphere} \\ Q & \text{if } p = O \end{cases}$$

(To check continuity of h or h^{-1} keep in mind that the single isolated point in each object is not "close" to any other point of the object.) Thus, X and Y are homeomorphic. However, a little intuition allows us to conclude that X and Y are *not* isotopic, since a continuous deformation of X into Y would certainly take O to Q and at some time during the motion, O would need to "move through" the sphere.

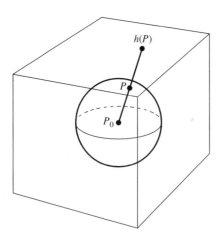

Figure 1.14 A visually defined homeomorphism.

This example has led us to a crucial point in our discussion of objects being "homeomorphic" versus being "isotopic." Previously we concluded that if two objects are isotopic, then they are also homeomorphic. Now it appears that the converse of this statement does not hold. In this book, we will not attempt to tackle the hard general problem of the precise relationship between these two notions, but we will be able to investigate that relationship a little further. In doing so, we will be forced to introduce some new ideas that will be important in other considerations as well.

It is important to point out that two objects being homeomorphic depends only on the objects themselves while two objects being isotopic depends also on the surrounding space in which the objects lie. Essentially it is the restricted nature of 3-space that gets in the way of finding a continuous deformation from object X to object Y in the previous example!

To motivate this further, let us consider the similar situation one dimension lower. Now let X be the subset of the x-y plane consisting of the unit circle along with the origin $(0, 0)$, and let Y consist of the unit circle along with the single point $(0, 2)$. Again, it is easy to write a rule for a homeomorphism from X onto Y. And again, if we restrict ourselves just to continuous deformations *in the plane,* we cannot find such a deformation that takes X onto Y. Of course, if we allow ourselves the extra freedom of working in 3-space, it becomes easy to just lift the origin up and move it over to coincide with the point $(0,2)$, as illustrated in Figure 1.15. We might say that X and Y are "isotopic in 3-space" but are *not* "isotopic in 2-space." Thus, to seek a continuous deformation in the previous example, we should investigate *four-dimensional space*.

Four-dimensional Euclidean space (*4-space*) is the set of all ordered quadruples (x, y, z, t) of real numbers. We think of these quadruples as points in some kind of "geometrical" space, but it is difficult to visualize this space. However, certain subsets are easier to think about. For example, we can identify the points $(x, y, z, 0)$ of 4-space with the points (x, y, z) of 3-space. So 3-space sits inside 4-space in a natural way, just as 2-space sits inside 3-space in a natural way. The Euclidean distance function for 4-space takes the distance between the points (x_1, y_1, z_1, t_1) and (x_2, y_2, z_2, t_2) to be $\sqrt{(x_1 - x_2)^2 + (y_1 - y_2)^2 + (z_1 - z_2)^2 + (t_1 - t_2)^2}$. Note that distances between points in 3-space are the same whether measured using the 3-space distance function or the 4-space distance function.

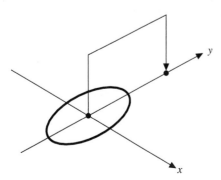

Figure 1.15 The freedom of 3-space frees a trapped point in the plane.

Now recall the objects X and Y of Example 1.18. These objects lie in 3-space, so we may consider them to lie in 4-space as well—all points now have 0 as a fourth coordinate. We can show that X and Y are "isotopic in 4-space" if we can continuously "lift" the origin point $O = (0, 0, 0, 0)$ in 4-space so that it never comes near any point $(x, y, z, 0)$ of the unit sphere and lands coincident with the point $Q = (0, 0, 2, 0)$. Begin by simply letting the fourth coordinate of O increase from 0 to 1. During this motion, the coordinates of the point are $(0, 0, 0, t)$, where $0 \leq t \leq 1$, so that the point gets further and further away (measured in 4-space distance) from the sphere. Next continue the motion of the point by letting the third coordinate increase from 0 to 2, and finally let the fourth coordinate decrease from 1 to 0, at which instant the point coincides with Q. During all three parts of this continuous motion, the distance between the point and the sphere was never less than the original distance of one unit, measured using the 4-space distance function!

This approach to considering objects as subsets of 4-space and deformations in 4-space will be very useful to us, especially in our study of nonorientable surfaces when the problem of apparent "self-intersection" will arise.

As far as the general notion of topological equivalence goes, **unless otherwise noted, throughout this book we will consider two objects to be topologically equivalent if they are homeomorphic.** Thus, topological equivalence will indeed reflect the "intrinsic" topological nature of objects themselves rather than their "extrinsic" nature, which involves the particular space in which they lie. In some cases, in order to see that two objects are homeomorphic we may instead observe that they are isotopic (in some Euclidean space).

• **Exercise 1.8**   As in Example 1.18, suppose that a point is trapped inside a sphere in 3-space. Explain in detail how the sphere can be continuously deformed in 4-space in such a way that the closed "southern hemisphere" (which includes the "equator") stays in 3-space while the open "northern hemisphere" moves out of 3-space, allowing the point to escape while staying in 3-space. Be careful to not disconnect the sphere during this continuous deformation in 4-space!

It is appropriate to conclude this section with a few more comments on the "ambient" nature of the notion of two objects being isotopic, which was mentioned in Section 1.1. We have learned, primarily through examples, that we must be careful to understand the space in which a continuous deformation is being performed. This reveals the "extrinsic" nature of the meaning of the word "isotopic." That space is called the "ambient" space of the deformation, and the formal, rigorous definition requires that the continuous deformation between two isotopic objects extend to a continuous deformation of *all* of the ambient space. In this book we will not dwell on this additional requirement. However, from a technical mathematical point of view, this "ambient" condition will be assumed as part of the meaning of "isotopic" whenever it is used.

1.4   CONSTRUCTING OBJECTS FROM PLANE MODELS

Section 1.1 included a description of the construction of paper models of a cylinder and a Möbius band. Many objects can be constructed in a similar fashion.

Such a general construction begins with a polygonal planar region homeomorphic to the closed unit disk. The actual construction involves gluing together various sets of edges, identified with labels, along their entire lengths in a direction specified on each edge by an arrowhead. (An actual gluing process may involve some amount of stretching of the planar region—even into 4-space—to make the associated edges meet!) The polygonal region, along with the various labels and arrowheads, is called a *plane model* for the object constructed.

Construction of objects using plane models will be important throughout much of this book, and a few interesting examples presented now should make the process clear.

EXAMPLE 1.19

Figure 1.16 shows simple square plane models for the cylinder and the Möbius band that illustrate the constructions already presented in Section 1.1. Note in both cases that the edges labeled a and c are not associated with any edges for gluing and in fact remain on the constructed objects to form their boundary sets.

EXAMPLE 1.20

In Figure 1.17, the square plane model shown indicates that top and bottom edges are to be glued together from left to right and left and right edges are to be glued together from top to bottom. Following this sequence of gluings step-by-step reveals that this is a plane model for the torus. Gluing together the top and bottom edges forms a cylinder that, when thought of as a rubber hose, can be bent so that the remaining edges (now the end circles of the hose) can be glued appropriately.

Imagine walking along the surface of the torus constructed from the above plane model, but instead of actually walking on the torus, walk on the square plane model. If you step off the top edge, you don't fall off. In fact you continue your trip, stepping across the bottom edge!

• **Exercise 1.9**   Jeffrey Weeks introduced a marvelous variation on a well-known game in his book *The Shape of Space* **[Weeks].** Imagine Tic-Tac-Toe played as usual on a square board of nine cells, but let the board actually be the plane model for a torus introduced in Example 1.20 so that, after gluing, the board can be thought of as covering the torus surface. For example, on the left board shown in Figure 1.18, player X has three X's in a row diagonally, even though this would not be the case in usual Tic-Tac-Toe.

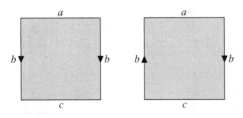

Figure 1.16 Plane models for a cylinder and a Möbius band.

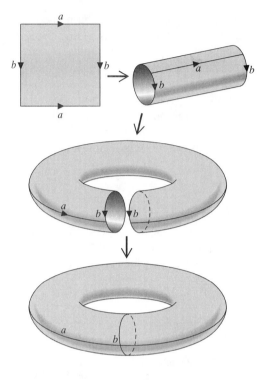

Figure 1.17 A torus formed from a plane model.

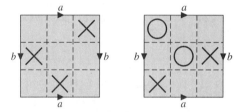

Figure 1.18 Tic-Tac-Toe on a torus.

Suppose that you are player X competing against player O on the right board shown in Figure 1.18 and it is your turn. You can block O's win by placing an X in the lower right square. Can you do better?

EXAMPLE 1.21

The gluings indicated in the plane model shown in Figure 1.19 produce a sphere. Note that before any gluings are made, the plane model is stretched first into the shape of a bag and then into a sphere with a slit.

EXAMPLE 1.22

This last example of the chapter concerns the plane model in Figure 1.20. Note that it closely resembles the plane model for the torus discussed in Example 1.20, except for the

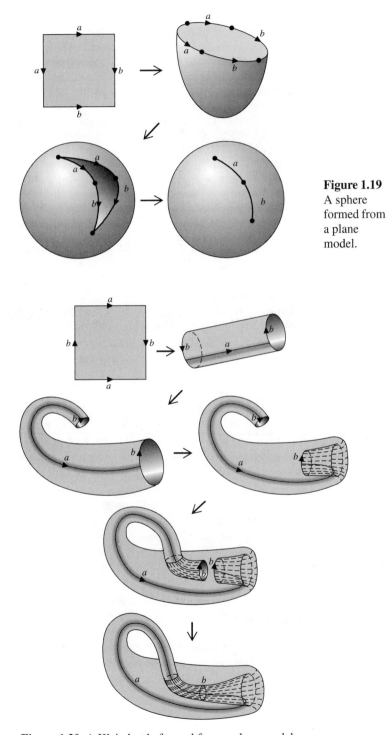

Figure 1.19
A sphere
formed from
a plane
model.

Figure 1.20 A Klein bottle formed from a plane model.

gluing direction arrowhead on the left edge. To see what object is produced from this model, we may follow the sequence of gluing steps in the figure. The first step, in which the top and bottom edges are glued together, again forms a cylindrical hose, but the second step cannot be completed as easily as before since the "end circles" do not meet properly when simply bent together. The proper orientation for the end circles to meet (shown in the figure) seems to require the hose to pass through itself. Such a motion can be performed in 4-space as follows. Divide the cylindrical hose into thirds and deform it in 4-space by continuously varying only the fourth ("t") coordinates of points so that at the end circles $t = 0$, from the end circles to the "third" dividing circles t increases linearly to $t = 1$, and throughout the middle third $t = 1$. The left end of the hose, say the portion corresponding to t-values between 0 and 1/2, may now be stretched and bent so as to pass through the middle third of the hose surface unobstructed in 4-space—although in our picture of the 3-space image we see a curve of "self-intersection!" The end circles are now properly oriented and can be glued together according to the specification of the plane model. The surface so constructed is the famous "closed one-sided surface" called the *Klein bottle*. It was first introduced by Felix Klein in 1882.

• **Exercise 1.10** Week's game of Tic-Tac-Toe on a torus could also be played on the plane model for a Klein bottle discussed in Example 1.22. Find an example of a winning "three-in-a-row" arrangement for Klein bottle Tic-Tac-Toe that would not win in either usual or torus Tic-Tac-Toe.

1.5 CHAPTER SUMMARY

In this chapter the mathematical subject of topology has been introduced from a visual point of view. Through an examination of many examples, sets of points in space have become the objects of our attention, and we have even been drawn to consider objects that lie in 4-dimensional Euclidean space! The notion of topological equivalence was motivated by the "extrinsic" similarity of objects—being *isotopic* in some space—but was formally developed as the "intrinsic" similarity of objects—being *homeomorphic*. Thus, the topological equivalence of two objects depends only on the objects themselves, not on the space in which they lie. Finally, a method for constructing some special objects by using plane models was developed and led to a common approach for describing some well-known "surfaces," including the sphere, the torus, and the Klein bottle.

In the next chapter we will turn our attention to a specific class of objects, namely *manifolds*, and their properties. Among the 1-dimensional manifolds will be "curves," and we will learn that, topologically speaking, curves are very easy to classify. Two-dimensional manifolds will be introduced, and we will investigate the types of 2-dimensional manifolds that may be constructed from plane models. Two-dimensional manifolds will remain a central focus in some later chapters as well.

1.6 SUPPLEMENTARY EXERCISES

Exercise S-1.1 The object shown on the left in Figure 1.21 is a two-holed torus on which a black has been painted as shown. Demonstrate, with a sequence of sketches,

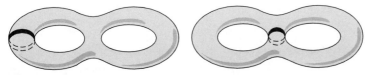

Figure 1.21 Surprisingly isotopic objects for Exercise S-1.1.

that this object can be continoulsy deformed *in* 3-*space* into the object shown on the right in the figure, with the painted band in an apparently completely different location.

Exercise S-1.2   Let X be an open disk along with a "half-open arc" of its boundary points, let Y be an open annulus along with one of its inner circle boundary points, let Z be an open annulus along with a "half-open arc" of its outer circle boundary points and one of its inner circle boundary points, and let D be an open disk. These four planar sets are shown in Figure 1.22. It is a fact that no two of these are homeomorphic.

(a) Mimic the bending of the half-open interval done in Example 1.16 to deform X in such a way as to suggestively define a continuous bijection from X onto Y. (*Hint:* Start by deforming X into a long rectangle that will play the role of the half-open interval in Example 1.16.)

(b) Repeat what you did in part (a) so as to suggestively define a continuous bijection from X onto Z.

(c) Show that there is a continuous bijection from Y onto D. Then also show that there is a continuous bijection from X onto D.

(d) In part (b), you showed there is a continuous bijection from X onto Z. Now show there is a continuous bijection from Z onto X. Does this contradict the fact that X and Z are not homeomorphic? Explain.

Exercise S-1.3   Consider the 3-sided plane model shown in Figure 1.23. The sides labeled "a" are to be glued together along their entire lengths, while the side labeled "b" is to be left alone and provide a boundary for the resulting object. There are two different ways in which the gluing of the "a" edges can be done to create objects. Actually cut out paper plane models for each of these objects, and tape together the "a" edges to build a model of each object. (*Hint:* The triangular plane models will need to be isosceles triangles, of course, but you may need to experiment a bit to determine appropriate

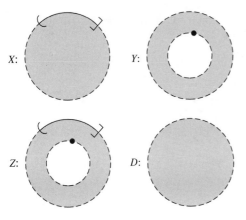

Figure 1.22 The four planar sets for Exercise S-1.2.

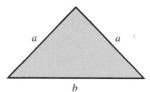

Figure 1.23 The triangular plane model for Exercise S-1.3.

base-to-height ratios.) These two objects are homeomorphic to objects introduced in this chapter. Identify them.

Exercise S-1.4   A seven-sided plane model is shown in Figure 1.24.

(a) After all the gluings are completed, how many points on the resulting object correspond to corner points of the seven-sided plane model?

(b) Show, in a sequence of sketches, successive deformations of the plane model, gluings, and so on, leading to a picture of the resulting object. Describe the resulting object in words.

Exercise S-1.5   A *knot* is a subset of 3-space which is homeomorphic to the unit circle in the plane. Most knots are not, however, isotopic in 3-space to the circle. But all knots are isotopic to the circle in 4-space. In other words, given the extra freedom of 4-space, all knots can be "unknotted!" Explain in detail, using words and pictures, how the knot shown in Figure 1.25 below can be unknotted in 4-space. This knot is called a "trefoil knot."

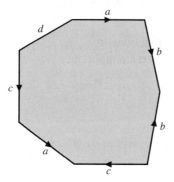

Figure 1.24 The plane model for Exercise S-1.4.

Figure 1.25 The trefoil knot for Exercise S-1.5.

Chapter 2

Manifolds

In this chapter we will introduce, through definitions and examples, certain types of objects in space called *manifolds*. Manifolds form a class of topological objects that includes knots (specific examples of one-dimensional manifolds) and surfaces (specific examples of two-dimensional manifolds), which we will investigate in detail in subsequent chapters. We will also briefly consider three-dimensional manifolds, which are objects of much interest in modern topology—primarily since they are involved in some challenging problems, many of which are still unresolved.

2.1 BASIC DEFINITIONS AND CURVES

When we were considering the continuity of functions in Chapter 1, the points in an object that were close to a given point played an important role. It turns out that sets of such points are fundamental in practically every topological consideration; they are called *neighborhoods*.

Definition

Let X be an object (i.e., a set of points in space) and let p be a point of X. A *basic neighborhood* of p is a set consisting of all points of X that lie strictly within a distance ε of p, where ε is some positive real number.

By a neighborhood of p we mean a subset of X that contains some basic neighborhood of p.

Note that a basic neighborhood of p contains p itself, but it is not guaranteed to contain any other points. There are always infinitely many points of space within ε of p, but only those that are *in the set X* belong to the basic neighborhood. Although we pointed out the meaning of the word *neighborhood* in the definition above for the sake of completeness, we will mostly consider basic neighborhoods of points and will even informally say "neighborhood" when referring to a basic neighborhood.

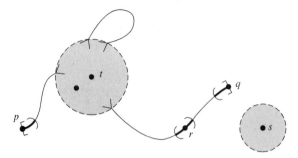

Figure 2.1 Various neighborhoods of points in a planar set.

EXAMPLE 2.1

Let X be the set of points in 2-space shown in Figure 2.1. It consists of an arc that is homeomorphic to the closed unit interval along with several other isolated points. The small basic neighborhoods of points of X are of three types: small basic neighborhoods of the "endpoints" of the arc (p and q in the figure) are homeomorphic to half-open intervals while "interior" points of the arc (such as point r) have small basic neighborhoods homeomorphic to open intervals; small basic neighborhoods of each of the isolated points (such as s) consist of just that isolated point itself. Note that we have used the reference "small" here to mean that we can find some (perhaps very small) positive number such that any basic neighborhood determined using ε smaller than that number is of the described type. Some "larger" basic neighborhoods of points of X could look a bit more complicated as the figure illustrates for the point labeled t where the neighborhood shown consists of two arcs and an isolated point.

It is easy to check that if an object X is a set of points in 2-space, then the neighborhoods of a point of X will be the same whether distances are computed under the assumption that X lies in 2-space or that X lies in 3-space. And similar remarks apply to an object in 3-space: its neighborhoods can be computed using either 3-space distance or 4-space distance.

EXAMPLE 2.2

Figure 2.2 shows a cylinder, a sphere, and a cube's surface. These are objects that we think of as "2-dimensional surfaces." Small basic neighborhoods of various points on these objects are shown in the figure. Note that all "interior" points of the cylinder and all points on the sphere or the cube's surface have small basic neighborhoods that are homeomorphic to the open unit disk in the plane—even the points on the corners or edges of the cube! However, points that lie on the boundary circles of the cylinder possess basic neighborhoods which are homeomorphic to the planar "half-disk" described by $x^2 + y^2 < 1$ and $y \geq 0$.

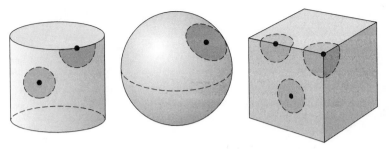

Figure 2.2 Neighborhoods of points on a cylinder, a sphere, and a cube's surface.

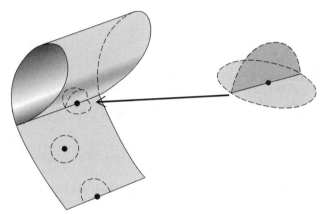

Figure 2.3 A point here has a 3-flap neighborhood.

EXAMPLE 2.3

The set of points in 3-space shown in Figure 2.3 might also be thought of as a "2-dimensional surface." Think of this set as being constructed from a sheet of paper by bending the top edge down to be glued to a horizontal line across the sheet. While some points of this set have small basic neighborhoods homeomorphic to planar open disks or planar "half-disks," the points that lie on the line to which the top edge was glued have small basic neighborhoods consisting of three "flaps" joined along a single edge.

• **Exercise 2.1**   Figure 2.4 shows a "flying saucer" object consisting of a spherical sur-face "body," a closed annulus "wing" whose inner circle is glued to the equator of the sphere, and a closed interval "antenna" with one endpoint glued to the north pole of the sphere. This object has points whose small basic neighborhoods are of various types. De-termine how many different types of small basic neighborhoods occur for the points of the flying saucer. Sketch a picture of each type of basic neighborhood found and give a description of each in words.

The examples we have seen illustrate an important idea about neighbor-hoods: the small basic neighborhoods of a point of an object describe the *local*

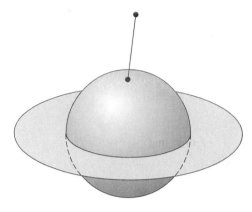

Figure 2.4 The "flying saucer" object of Exercise 2.1.

nature of the object near the point. An object will be "locally like the real line" near a point if that point has a basic neighborhood that is homeomorphic to an open interval while it will be "locally like the plane" near a point if that point has a basic neighborhood that is homeomorphic to an open planar disk. These ideas lead us to the following definition.

Definition

A set of points in space is a 1-*dimensional manifold* if each of its points has a neighborhood that is homeomorphic to an open interval of the real line.

A set of points in space is a 2-*dimensional manifold* if each of its points has a neighborhood that is homeomorphic to an open planar disk.

The reader now might wish to formulate the definition of 3-*dimensional manifold*. We will postpone the official discussion until section 2.4.

The simplest examples of 1-dimensional manifolds are 1-space itself and open intervals in 1-space. Unions of open intervals provide other examples of subsets of 1-space that are 1-dimensional manifolds, so 1-dimensional manifolds need not have just one component. But 1-dimensional manifolds need not be subsets of 1-space. Any set of points in space that is homeomorphic to a circle is also a 1-dimensional manifold. We must be cautious and realize, however, that there are some sets that seem "1-dimensional" that are not 1-dimensional manifolds. For instance, a closed interval of the real line fails to be a 1-dimenssional manifold because its endpoints fail to have neighborhoods that are homeomorphic to an open interval.

Likewise, the simplest examples of 2-dimensional manifolds lie in the plane: these are 2-space itself and subsets of 2-space that are homeomorphic to a planar open disk. There are many more intricate subsets of 2-space that are 2-dimensional manifolds, such as an "open" annulus (containing the points strictly between two concentric circles) or a union of open disks. Both the sphere and the torus are examples of subsets of 3-space that are 2-dimensional manifolds, and the Klein bottle is a subset of 4-space that is a 2-dimensional manifold! However, the cylinder and the Möbius band (considered to include their boundaries) are not 2-dimensional manifolds since points on the boundaries fail to have neighborhoods that are homeomorphic to planar open disks.

One of our goals in investigating manifolds will be to describe what they all look like "up to homeomorphism." This task, that is sometimes called "classification," can be simplified a bit if we restrict our attention to manifolds that satisfy several further properties. The properties that we next consider are somewhat natural and we will treat them intuitively, although in a more technical and rigorous approach they would require a much deeper mathematical background than we expect of most readers of this book.

Definition

A set X of points in space is

(a) *connected* if X consists of just one component;
(b) *bounded* if there is an upper bound for distances between pairs of points of X;
(c) *closed* if X contains all points in space that are limits of sequences of points of X; and
(d) *compact* if X is both bounded and closed.

In most of our discussion of manifolds in this book, we will restrict our attention to those that are connected and compact. We have already mentioned that manifolds may contain several components; so by assuming connectedness of those we study, we are investigating manifolds "one piece at a time." The compactness (bounded plus closed) assumption actually provides some technical help for completing some of the mathematical details of the classification process. A set with the property of being bounded is contained in some finite-sized region in space, so it doesn't extend infinitely far out into space; if the set is also closed, then any points of space that seem like they should be "boundary points" of the set are actually in the set. Before looking at some motivating examples, we must point out that our definition of compactness can be reformulated in several different forms, some of which are useful in defining compactness in more general topological settings. However, the definition we are using is entirely adequate for our purposes in this book.

EXAMPLE 2.4

Observe that 1-space, the real line, is homeomorphic to the open interval $(-1, 1)$. (One homeomorphism from $(-1, 1)$ to the real line is provided by a continuous function familiar from calculus: $h(x) = \tan(\pi x/2)$. Note that the inverse function from the real line to $(-1, 1)$ is given by the rule $h^{-1}(x) = \frac{2}{\pi}\tan^{-1}(x)$, which certainly defines a continuous function also.) Thus, topologically speaking, 1-space and the interval $(-1, 1)$—or *any* open interval, for that matter—are the same 1-dimensional manifold. Let us think about the properties just introduced. Clearly 1-space and the interval $(-1, 1)$ are both connected. Now 1-space is closed (since it clearly contains all its "potential" boundary points) but it is certainly not bounded. On the other hand, $(-1, 1)$ is clearly bounded but is certainly not closed (since it does *not* contain its "potential" boundary points -1 and 1). So neither 1-space nor $(-1, 1)$ is compact, and each fails to be compact for a different reason!

EXAMPLE 2.5

The closed interval $[0, 1]$ is certainly connected, and it is also compact. Recall, however, that $[0, 1]$ is not a 1-dimensional manifold.

EXAMPLE 2.6

Note that 2-space and the open unit disk in the plane are homeomorphic. (The reader is challenged to provide an explicit homeomorphism to demonstrate this.) Both 2-space and the open disk are connected, but neither is compact: again, one is bounded but not closed, and the other is closed but not bounded.

EXAMPLE 2.7

We can easily check that the unit sphere is a connected and compact 2-dimensional manifold. The surface of a cube, which we know is homeomorphic to the sphere, can also be checked for these properties.

• **Exercise 2.2**   For each of the sets below, describe the various types of "small" basic neighborhoods of its points. For each, say whether the set is a 1-dimensional manifold, a 2-dimensional manifold, or neither. For each, also say whether the set is connected and whether it is compact. In each part, a sketch will be very helpful!

(a) $\{(x,y): y^2 > x^2 + 1\}$
(b) $\{(x,y,z): x^2 + y^2 + z^2 = 4 \text{ and } x \le 1\}$
(c) $\{(x,y): x = \sin(t) \text{ and } y = \sin(2t) \text{ where } 0 \le t \le 2\pi\}$

The examples that we have seen just start to suggest some important facts, which we will not prove but can certainly understand intuitively. If two sets are homeomorphic, then it need not be true that both of them are bounded or that both of them are closed. However, the properties of connectedness and compactness are "topological properties" in the sense that they are preserved by homeomorphisms.

> **Theorem**
>
> For two sets X and Y of points in space, assume that there is a homeomorphism from X to Y.
>
> **(a)** If X is connected, then Y is also connected.
> **(b)** If X is compact, then Y is also compact.

Actually, the conclusions of this theorem follow from the weaker hypothesis that there is a continuous function from X onto Y. Thus, in particular, a continuous image of a connected object will be connected itself. This result generalizes the intermediate value theorem, which the reader may recall from calculus.

To get a feel of what "classification" is all about, we will complete this section with a discussion of what all the connected 1-dimensional manifolds look like topologically. At this time it is somewhat natural to introduce the word "curve" into our discussion.

> **Definition**
>
> A *curve* is a connected 1-dimensional manifold.

Among the curves are any sets of points in space that are homeomorphic to the real line $(-\infty, \infty)$, but these all fail to be compact since they are either not bounded, not closed, or both. On the other hand, the unit circle in the plane is an example of a compact curve, as is any set homeomorphic to it. Our objective of "classifying" the sets of points in space that are curves amounts to finding a "short list" of specific examples of curves such that any curve is homeomorphic to one of those on the "short list." A little experimenting at drawing different curves will likely lead the reader to conjecture that the two examples mentioned earlier in this paragraph—the real line and the unit circle—actually provide the required "short list." Of course, one of these—the unit circle—represents all the compact curves while the real line represents all those curves that are not compact. This conclusion is stated formally below.

> **Theorem**
>
> *The Classification Theorem for Curves*
>
> **(a)** Any compact curve is homeomorphic to the unit circle.
> **(b)** Any noncompact curve is homeomorphic to the real line.

This result, being a "Theorem," can of course be proven. We will not pursue the proof, which requires delicate use of the compactness notion and some other technical general topological properties. It also involves mathematical induction, a proof technique that will be introduced in the next chapter. There we will undertake a quite motivating and instructive approach to proving the interesting classification theorem for compact, connected 2-dimensional manifolds. We will begin a preliminary discussion of these objects in the next section.

2.2 SURFACES

A typical intuitive exposition on topology that might be offered to nonmathematicians cites its role as a "rubber sheet geometry," in which surfaces illustrate the basic elementary ideas quite well. It is also true that properties of surfaces were of much interest to mathematicians when the early ideas of topology were being developed and are still of interest, especially in motivating the study of higher-dimensional manifolds and more general topological spaces. Here is an official definition.

> **Definition**
>
> A surface is a connected 2-dimensional manifold.

The whole x-y plane, or any object homeomorphic to 2-space, is a simple example of a surface. In multivariate calculus, students consider the graph in 3-space of the *circular paraboloid* with equation $z = x^2 + y^2$, graphed over the whole plane, which is a surface homeomorphic to 2-space. The same students likely consider the graph of $z^2 = x^2 + y^2 + 1$, graphed over the whole plane; this graph—*a hyperboloid of two sheets*—is not a surface since it is not connected. In Figure 2.5 a portion of the paraboloid is shown on the left, and a portion of the hyperboloid (including both components) is shown on the right. An "open" cylinder (without its top and bottom circular boundaries) is a surface, while the regular cylinder (which includes those boundaries) is not a surface since it fails to be a 2-dimensional manifold.

Every example of a surface mentioned in the last paragraph either fails to be closed or fails to be bounded; so none of them is compact. The classification of surfaces that we will undertake in the next chapter will be restricted to compact surfaces. It was pointed out in section 2.1 that the sphere is a compact and connected 2-dimensional manifold, so it is a simple example of a compact surface. After some inspection, the reader will agree that the torus is also a compact surface, and it is one that is not homeomorphic to the sphere. It will be our goal in the next chapter to completely determine all compact surfaces by giving a list of easily described compact surfaces so that any compact surface will be homeomorphic to one on the list.

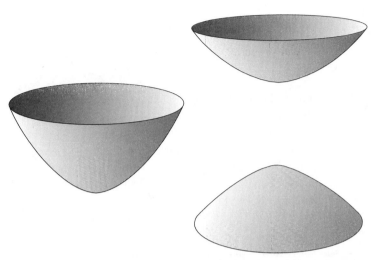

Figure 2.5 Portions of the graphs of a circular paraboloid, on the left, and a hyperboloid of two sheets, on the right.

The very word "surface" conjures up lots of examples in the sense that it suggests surface *of* some kind of object. Simply put, the compact surfaces we encounter in 3-space are just surfaces covering ordinary solid objects: the surface of a solid ball (a *sphere*), the surface of a solid donut (a *torus*), the surface of a solid two-holed donut (a *two-holed torus*), or the surface of a sheet of perforated postage stamps (the sheet—with some actual thickness—is the solid object; its "surface" is the connected 2-dimensional manifold). Note that here the phrase "ordinary solid object" really means what it says. A hollowed out object won't work. For example, the object consisting of a hollowed out ball (a spherical shell, if you wish) would be covered by a zero-thickness set that is a compact 2-dimensional manifold but not a surface since it is not connected; it has two components, an "inside" sphere and an "outside" sphere.

But there are some other 2-dimensional manifolds that do not occur as subsets of 3-space. For example, when we constructed the Klein bottle from a square plane model in Chapter 1, we glued together pairs of opposite edges of the square, forming an unbroken surface. In fact—as we will see for some other objects constructed from plane models in the next section—the Klein bottle is an example of a compact surface, and it is one that we have been able to describe (without self-intersections) in 4-space. On the other hand, the sphere, the torus, the two-holed torus, and any "surface of an ordinary solid object" all appear as surfaces in 3-space. It is of interest to know which compact surfaces can "live" completely in 3-space and which require 4-space to exist without self-intersections.

To address this issue, we need to discuss the notion of "orientability" of objects in space. The general idea, phrased very intuitively here, is that an orientable object should contain no "orientation-reversing paths." The Möbius band, with which we are already familiar, arises as a helpful tool, even though it is *not* a surface according to our definition. In the Möbius band there is a path that we,

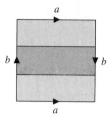

Figure 2.6 The Klein bottle contains a copy of the Möbius band.

as completely flat individuals trapped in the surface, could travel along and return to our starting point left and right reversed. Figure 2.6 shows, on the plane model we use to construct the Klein bottle, a horizontal strip that corresponds to a copy of the Möbius band inside the Klein bottle. We will use this idea to formally define *orientability* and *nonorientability* for compact surfaces.

Definition

A compact surface is said to be *nonorientable* if it contains a subset that is homeomorphic to the Möbius band; otherwise, it is called *orientable*.

Nonorientable compact surfaces are those that exist only in 4-space without self-intersections. It is the orientable surfaces that are the "surfaces of ordinary solid objects" in the sense of our three-dimensional experience. As the reader might guess, when we state the classification theorem for compact surfaces in the next chapter, we will need to treat the orientable and nonorientable cases separately.

• **Exercise 2.3** (a) Drilling a cylindrical hole through a hollow spherical shell of some positive thickness produces a solid object whose surface is a connected compact surface. Demonstrate, with a sequence of sketches, that this surface can be continuously deformed into a sphere.

(b) Make a conjecture about the compact surface obtained as the surface of a hollow spherical shell with two cylindrical drill holes. Draw a sequence of sketches to support your conjecture.

2.3 COMPACT SURFACES AND PLANE MODELS

In Chapter 1, we were able to construct some compact surfaces—as well as some compact objects that are not surfaces—from plane models. This is a very important procedure that we will refine and use as we continue to study compact surfaces. Recall that a plane model for an object consists of a polygonal region in the plane that is homeomorphic to the closed unit disk and on which each edge has a label and any edge with a label used more than once is assigned a direction indicated by an arrow. The object is constructed by gluing together edges of the plane model that share the same label along their entire lengths in the directions specified by their arrows. The plane model itself, being homeomorphic to the closed unit disk, is connected and is also compact (i.e., closed and bounded). Since the specified gluing process does not break apart the initial plane model, the object that results from the construction is clearly connected. Also the process of gluing together the appropriate edges can take place in a finite-sized region in space (perhaps, if necessary, in 4-space), so the object constructed in this fashion is bounded itself. Further, the gluing process does not introduce any possibility of new "potential boundary points" for the constructed object that were not among the points of the original plane model, so the constructed object is itself also closed. Therefore, such objects automatically have several desirable properties.

Remark

Any object constructed from a plane model is automatically connected and compact.

This fact can also be seen from properties of continuous functions. The gluing process assigns to each point p of the plane model a specific point $f(p)$ of the object constructed; $f(p)$ is the point of the object on which p falls after the construction is completed. This observation produces a continuous function f with domain being the plane model and range being the constructed object. It follows from remarks made in section 2.1 that the object must be connected and compact since it is the continuous image of the plane model, which is connected and compact.

We saw in Chapter 1 that the cylinder and Möbius band can both be constructed from plane models. Certainly they both are connected and compact (consistent with the above conclusion), but neither is a 2-dimensional manifold. If the reader reexamines the plane models of Figure 1.16, it will be clear that points on the so-constructed cylinder or Möbius band that fail to have neighborhoods homeomorphic to open disks are precisely those points of the plane model that lie on edges whose label is not shared with any other edge. Such edges, which were boundary edges of the plane model, remain as boundary edges of the constructed object. Thus, if we wish to construct a 2-dimensional manifold from a plane model, each edge label must be shared by at least two edges of the plane model. On the other hand, if an edge label is shared by more than two edges, then the points of the constructed object corresponding to points of such edges will have "multiple-flap" basic neighborhoods with three or more "flaps" and the object will fail to be a 2-dimensional manifold. (A point with "three-flap" basic neighborhoods was seen in the object discussed in Example 2.3.)

When an object results from the gluing process associated with a plane model, we will say that the plane model *represents* the object. We now can state precisely which plane models represent compact surfaces.

Theorem

A plane model represents a compact surface if and only if it has an even number, say $2n$, of edges identified in pairs by n different edge labels.

EXAMPLE 2.8

We have previously seen plane models representing the sphere, the torus, and the Klein bottle, and they are shown again in Figure 2.7. Note that each of these plane models has four edges and involves precisely two edge labels—the letters a and b—each used twice.

It is important to realize that a compact surface can be represented by many different plane models. Often a given compact surface can be "sliced up" and "spread out" into a plane model, and different such slicing-and-spreading out schemes can produce different plane models.

EXAMPLE 2.9

Figure 2.8 shows how two new plane models for the sphere may be obtained. The first is obtained by making a single slice on the sphere; this produces a 2-gon plane model whose two edges share the label a. The second plane model for the sphere is obtained by making seven slices on the sphere, each corresponding to a particular edge of a cube's surface. The resulting plane model is a 14-gon.

EXAMPLE 2.10

Figure 2.9 shows how a "two-holed torus," which is a compact surface, can be sliced along the curves labeled a, b, c, and d and spread out into an 8-gon plane model representing the two-holed torus.

Complicated compact surfaces are difficult to visualize (and even more difficult to draw) in their actual space settings—especially if they are situated

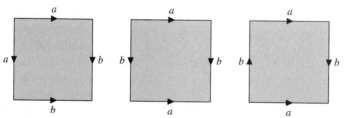

Figure 2.7 From left to right, plane models for the sphere, torus, and Klein bottle.

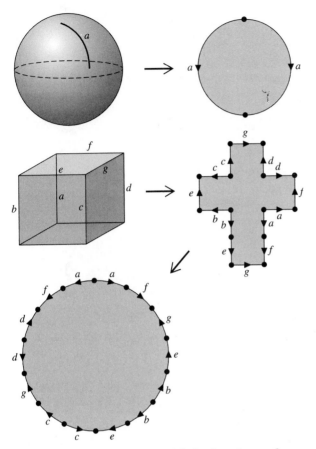

Figure 2.8 Two more plane models for the sphere: a 2-gon and a 14-gon.

in 4-space. Therefore, it will be very convenient to instead work on plane models that represent compact surfaces of interest and are easy to draw in the plane. But this poses a question: "Does every compact surface have a plane model that represents it?" Fortunately, the answer is yes. This follows from a deeper result called the *Triangulation Theorem,* a name we will borrow for our purposes.

Theorem

The Triangulation Theorem

Every compact surface can be sliced along finitely many curves and spread out into a plane model that represents it.

The word "triangulation" is associated with this result because, in an ultimately simplistic way, the resulting plane model can be further cut apart into

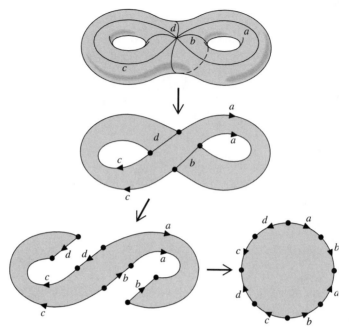

Figure 2.9 Obtaining an 8-gon plane model for a two-holed torus.

triangles that could be pieced back together to form the compact surface. The involved proof of this fact, which we will not attempt to describe, was first presented rigorously in the 1920s by a mathematician named Rado.

It turns out that there is a simple way to determine from a plane model the orientability or nonorientability of the compact surface it represents. If we look one more time at the now familiar plane model for the Klein bottle in Figure 2.6, the copy of the Möbius band shown in it is a strip connecting a portion of the one clockwise directed edge labeled *b* to the other clockwise directed edge labeled *b*. In fact, it is precisely this occurrence of two similarly directed and identified edges that determines which plane models for compact surfaces correspond to nonorientable compact surfaces.

Remark

Consider a plane model representing a compact surface *X*.

(a) *X* is orientable if and only if the two edges sharing each edge label occur with opposite directions, one clockwise and one counter-clockwise.

(b) *X* is nonorientable if and only if, for at least one edge label, the two edges sharing that label occur with the same direction, either both clockwise or both counter-clockwise.

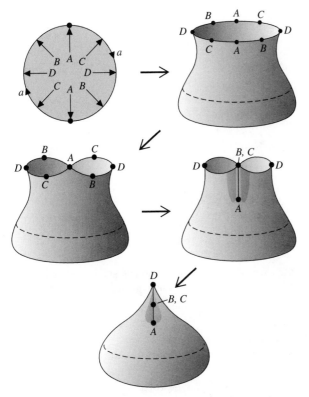

Figure 2.10 Obtaining the projective plane from a plane model.

The reader has probably guessed that there must be many more nonorientable compact surfaces besides the Klein bottle. In the following example, we introduce one such surface that will play an important role in the classification of nonorientable compact surfaces in the next chapter.

EXAMPLE 2.11

The *projective plane* is the compact surface represented by the 2-gon plane model shown in Figure 2.10, with two edges both labeled a and both directed clockwise. The so-described gluing simply dictates that points on the boundary of the plane model that are on opposite ends of each diameter should be identified as one point of the projective plane as shown for the points marked A through D. While this may be a simple description of the plane model, it is not so easy to visualize what this nonorientable compact surface looks like in 4-space. In the sequence of steps in the figure, we can follow the "stitching" of the a edges, starting with the points labeled A. As we continue sewing up the "seams," it becomes clear that one seam (the one starting at A and continuing through B) gets in the way of the other seam (the one starting at A and continuing through C). But the extra freedom of 4-space allows the seams to be completed, finally ending at point D, with only an apparent line of self-intersection in 3-space.

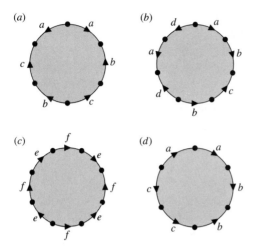

Figure 2.11 The four plane models for Exercise 2.4.

• **Exercise 2.4**   Decide whether each of the four plane models shown in Figure 2.11 represents a compact surface and, if it does, also determine whether the compact surface it represents is orientable or nonorientable.

Among the compact surfaces introduced thus far are four that are essential for our further investigation: the sphere, the torus, the Klein bottle, and the projective plane. We will, in the next chapter, be able to use these to describe all compact surfaces. These four surfaces are important enough to deserve their own mathematical symbols: S (the sphere), T (the torus), K (the Klein bottle), and P (the projective plane). The impatient reader, who cannot wait to see how these four objects combine to form all compact surfaces, may immediately proceed to Chapter 3. However, that reader will miss a tantalizing introduction to 3-dimensional manifolds in the next section of this chapter.

2.4   THREE-DIMENSIONAL MANIFOLDS

We live in a 3-dimensional universe that we intuitively think of as Euclidean 3-space. Small basic neighborhoods of all points in 3-space are alike: they are solid open balls of positive radius centered at some point, and such balls are all homeomorphic to the open unit ball $\{(x, y, z): x^2 + y^2 + z^2 < 1\}$. Given our experience with 1-dimensional manifolds and 2-dimensional manifolds, it is now easy to formulate the definition of a 3-dimensional manifold in such a way that 3-space is one of the easiest examples of such an object.

Definition

A set of points in space is a 3-*dimensional manifold* if each of its points has a neighborhood that is homeomorphic to the open unit ball $\{(x, y, z): x^2 + y^2 + z^2 < 1\}$.

Certainly 3-space and some other subsets of 3-space, such as open balls of finite positive radius or unions of such balls, satisfy this definition and thus are 3-dimensional manifolds. It is somewhat harder to visualize examples of 3-dimensional manifolds that possess some of the other properties we considered for 1-dimensional or 2-dimensional manifolds. We might be interested, for example, in getting a feel of what some compact connected 3-dimensional manifolds might look like. In fact, some physicists are very interested in studying precisely such objects, which they believe might be useful in offering an alternative model of the universe—a model that portrays the universe as being of "finite size" and "closing in on itself" in some sense. In this section, we will look at just a few interesting examples that should be fun, even if a bit challenging to visualize.

One of the simplest ways to construct some examples of compact connected 3-dimensional manifolds is to start with a three-dimensional version of a plane model consisting of a solid, closed polyhedron in 3-space (which would be homeomorphic to a solid *closed* ball) and gluing together the faces of the polyhedron in some specified way.

EXAMPLE 2.12

The 3-*dimensional torus* is obtained by gluing together opposite faces of a solid cube in 3-space, as shown for one of the three such pairs of faces in Figure 2.12. Although it is not easy to imagine the appearance of the resulting object, we can understand it better by considering motion through the object. If a mass inside the cube moves along a path that intersects a face at a certain point, then—in the 3-dimensional torus—the mass reappears entering the cube through the opposite face at the corresponding point.

Note that the closed solid cube we started with in Example 2.12 is not a 3-dimensional manifold, since any point on its surface has small basic neighborhoods that are homeomorphic to half of a ball without the hemispherical surface but with the open equatorial disk surface. When points on opposite faces are identified as one point in the gluing process, two such half-balls combine to form a neighborhood of that point in the 3-dimensional torus homeomorphic to an open ball. Similarly, four basic neighborhoods of any four identified points on edges of the cube come together to form an open ball neighborhood in the

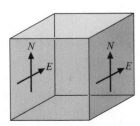

Figure 2.12 Gluing opposite faces produces the 3-dimensional torus.

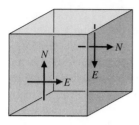

Figure 2.13 Gluing opposite faces with a quarter turn.

3-dimensional torus, as do eight basic neighborhoods of the eight corner points that all group together as *just one point* of the 3-dimensional torus.

• **Exercise 2.5** A 3-dimensional manifold may be constructed from a solid closed cube by gluing each face to the face opposite it with a one-quarter clockwise turn, as shown for two sides of the cube in Figure 2.13. How do the cube's corners group together when these gluings are made? (Recall that when the 3-dimensional torus was formed from a cube in a similar way, all eight of the corners came together as just one point of the 3-dimensional manifold.) This compact connected 3-dimensional manifold is referred to as the *quaternionic manifold* in [**Weeks**] where it serves as an example in the discussion of geometries on 3-dimensional manifolds.

EXAMPLE 2.13

A 3-dimensional sphere is actually quite easy to define. Think first about a 1-dimensional "sphere" (a circle). It may be represented as the unit circle in the plane—the set of all points in the plane that are exactly one unit distant from the origin. Similarly, a 2-dimensional sphere is represented by the unit sphere in 3-space—all the points in 3-space at a distance of exactly one unit from the origin. We will define the 3-*dimensional unit sphere* to be precisely what the reader has certainly predicted from the lower dimensional cases: it's the set of points in 4-space that are exactly a distance of one unit from the origin (0, 0, 0, 0). It is challenging to visualize, in this set, a small basic neighborhood homeomorphic to an open 3-dimensional ball for each of its points. However, a different intuitive—although abstract—approach to constructing the 3-dimensional sphere is helpful in revealing that it is indeed a 3-dimensional manifold. Again, begin by looking at the lower dimensional analogue. The 2-dimensional unit sphere looks like two closed planar disks (the upper and lower hemispheres) glued together along their circular boundaries so that, on the sphere, the two disks meet in a set homeomorphic to a 1-dimensional sphere (really a circle—the equator). Surprising as it seems, the 3-dimensional sphere can be visualized as two copies of a solid 3-dimensional ball whose boundary surfaces are glued together. So, these solid balls play the roles of the upper and lower "hemispheres" and the "equator" is homeomorphic to a 2-dimensional sphere!

EXAMPLE 2.14

For one last example, we will construct a 3-dimensional manifold that is somewhat like the Klein bottle. Start with the same solid 3-dimensional cube we used in constructing the 3-dimensional torus, and again glue the top to the bottom and the left face to the right face just as before. But this time, glue the front face to the rear face with a left-right re-

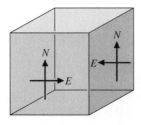

Figure 2.14 Obtaining a non-orientable 3-dimensional manifold.

versal as shown in Figure 2.14. The resulting 3-dimensional manifold is nonorientable in the sense that following a path forward through the front face brings you back into the cube through the rear face left-right reversed. This is an example of an "orientation reversing path" in this 3-dimensional manifold.

The investigation of 3-dimensional (and higher dimensional) manifolds is an important part of current research in topology. In this section we have barely glimpsed a few examples that illustrate the visual challenge of understanding such objects. A reader who wishes to learn more about the notions introduced here is encouraged to begin doing so by reading Jeffrey Weeks' wonderful book *The Shape of Space* [**Weeks**].

2.5   CHAPTER SUMMARY

In this chapter manifolds have been introduced as objects in space that, topologically speaking, look alike near all points. More precisely, all points of a manifold possess neighborhoods homeomorphic to a specific set: an open interval of the real line in the case of a 1-dimensional manifold and an open planar disk in the case of a 2-dimensional manifold. The properties of connectedness and compactness for objects were introduced, and curves and surfaces were defined as connected 1-dimensional and 2-dimensional manifolds, respectively. The classification theorem for curves specified two particular curves, the compact unit circle and the noncompact real line, which topologically represent all curves in the sense that any curve is homeomorphic to one of them. Although we stated this theorem without proving it rigorously, it served to illustrate the general idea of "classification" of topological objects. Properties of objects that can be constructed from plane models were investigated. In particular we determined conditions under which a plane model produces a compact surface as well as conditions that determine the orientability of such a compact surface. Finally, we took a quick trip into the realm of 3-dimensional manifolds and examined some interesting compact examples among them.

In Chapter 3 we will undertake a proof of a classification theorem for compact surfaces. The "algebraic" tools we will develop to accomplish that task rely heavily on plane model constructions, and four familiar compact surfaces—the sphere, torus, Klein bottle, and projective plane—will serve as building blocks in our work.

2.6 SUPPLEMENTARY EXERCISES

Exercise S-2.1 Consider the subset X of the real line whose members are 0 and the numbers of the form $1/n$, where n is a positive integer.

(a) Describe the types of small basic neighborhoods possessed by points of X.

(b) Is X connected? Why or why not?

(c) What subsets of X are connected? Explain your answer.

(d) Is X compact? Why or why not?

(e) What subsets of X are compact? Explain your answer.

Exercise S-2.2 **(a)** Give an example of two disjoint subsets of the x-y plane, each of which is homeomorphic to the open unit interval (0, 1) and for which the union of the two subsets is not a 1-dimensional manifold. Explain why such sets cannot both be chosen to be subsets of the real line.

(b) Give an example of two disjoint subsets of 3-space, each of which is homeomorphic to the open unit disk and for which the union of the two subsets is not a 2-dimensional manifold. Explain why such sets cannot both be chosen to be subsets of the x-y plane.

Exercise S-2.3 Following a procedure similar to that used in Example 2.10 (illustrated in Figure 2.9), draw some curves on a sketch of a three-holed torus and show how, by slicing along the curves and spreading out the surface, you can obtain a plane model for the three-holed torus.

Exercise S-2.4 A 2-dimensional *manifold with boundary* is a set in space in which each point has a neighborhood homeomorphic to either an open planar disk or a "half disk" such as $\{(x, y) : x^2 + y^2 < 1 \text{ and } y \geq 0\}$. Note that any 2-dimensional manifold is a 2-dimensional manifold with boundary.

(a) Give some examples of 2-dimensional manifolds with boundary that are not 2-dimensional manifolds. Include some examples that are connected, some that are not connected, some that are compact, and some that are not compact.

(b) Determine what type of plane model will produce a 2-dimensional manifold with boundary.

Exercise S-2.5 *Three-dimensional projective space* is the 3-dimensional manifold obtained by starting with a solid closed 3-dimensional ball and identifying points on the ball's surface that are at opposite ends of a diameter. Is this 3-dimensional manifold non-orientable in the sense that it admits an orientation reversing path? Explain your answer.

Chapter 3

Classification of Compact Surfaces

In this chapter we will develop a quite elegant description of all compact surfaces by showing that each such object is homeomorphic to a member of a collection of particularly simple (or at least simply described) compact surfaces. Those special surfaces, which we will call *normal form* surfaces, occur as certain combinations of some elementary compact surfaces we have already studied, namely, *S*, the sphere; *T*, the torus; *K*, the Klein bottle; and *P*, the projective plane. While developing this result, we will see that most of our work with surfaces—thought of as (sometimes very complicated looking) objects in space—can be simplified by instead treating them as certain types of "algebraic" objects.

3.1 CONNECTED SUMS: NEW SURFACES FROM OLD

It has probably occurred to some readers that a two-holed torus must, in some sense, be two one-holed tori combined. That is precisely right. Imagine two disjoint homeomorphic copies *X* and *Y* of the torus. Remove a small open disk from each of *X* and *Y*, leaving the boundary circles on the tori. What remain now of *X* and *Y* are no longer surfaces since, having boundary edges, they are no longer 2-dimensional manifolds. Now we wish to "connect" these two objects by gluing together the boundary circle on *X* and the boundary circle on *Y*. This gluing process might be seen more easily if we, to whatever extent necessary, deform *X* and *Y* so that the circles appear on the ends of tubes projecting from each surface. (See Figure 3.1.) When a point on the boundary circle of *X* is glued to a point on the boundary circle of *Y*—forming a single point—half-disk basic neighborhoods of each of the two points combine to produce an open disk neighborhood of the new single point. Thus, the object constructed in this way is indeed a 2-dimensional manifold and in fact a compact surface. And further, as the figure suggests, that object is (at least, homeomorphic to) a two-holed torus.

Figure 3.1 Forming the connected sum of two tori.

The construction we have just illustrated is called the *connected sum* construction, and it allows us to combine any two compact surfaces to obtain another compact surface:

Definition

If X and Y are two compact surfaces, then the *connected sum* of X and Y is the compact surface constructed by the following steps:

(1) Remove a small open disk from each of the sets X and Y, leaving the boundary circles on each of the surfaces;
(2) Glue together the boundary circles to form the connected sum.
   The connected sum of X and Y is denoted by $X\#Y$.

Several observations about connected sums should be made now. It does not matter *where* on either surface the open disk is removed; if Y gets "attached" to X at one location on X, then $X\#Y$ can be continuously deformed as necessary so that Y appears to be attached to X at another location. Also, $\#$—the connected sum "operation"—is defined as a "binary" operation in the sense that it combines just two objects at a time. However, applied repeatedly, $\#$ can produce a connected sum of three, four, or any finite number of compact surfaces. One would intuitively expect that forming the connected sum of $X\#Y$ and Z would produce a surface homeomorphic to the connected sum of X and $Y\#Z$, and that is the case. In other words, the connected sum operation $\#$ satisfies an *associative* property. It is also intuitively clear that a connected sum $X\#Y$ is homeomorphic to $Y\#X$, so that is also *commutative*. Therefore, we can form connected sums of finitely many compact surfaces without worrying about ambiguity due to the way in which the surfaces are ordered or grouped.

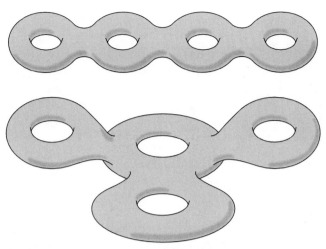

Figure 3.2 Two different views of a four-holed torus.

EXAMPLE 3.1

According to the last mentioned remark, we can construct the connected sum of n tori, where n is any positive integer. We will use the notation nT to denote the resulting connected sum. Figure 3.2 shows two objects, both of which are homeomorphic to $4T$, although the way their individual tori are attached appears quite different.

• **Exercise 3.1**   Draw a sequence of sketches showing how one of the two views of 4T shown in Figure 3.2 can be continuously deformed into the other view.

EXAMPLE 3.2

What does the connected sum of a compact surface X and the sphere S look like? Figure 3.3 shows quite clearly that it looks just like X itself! In other words, $X\#S$ is homeomorphic to X. Of course, a little thought convinces us this must be the case: a sphere with an open disk removed is homeomorphic to a closed planar disk which, in the connected sum construction, simply patches up the hole which was made in the surface X.

We may now preview the main result of this chapter. Just as we have already used nT to denote the connected sum of n tori, we will also use mP to denote the connected sum of m homeomorphic copies of the projective plane. By *normal form surface* we will mean any one of the compact surfaces S, nT ($n \geq 1$), or mP ($m \geq 1$). We will be showing that any compact surface is homeomorphic to a normal form surface. Of course, among the normal form surfaces, S and nT ($n \geq 1$) are orientable and mP ($m \geq 1$) are nonorientable.

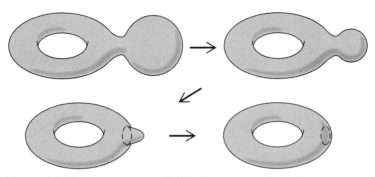

Figure 3.3 The connected sum $X\#S$ is homeomorphic to X.

In this list of normal form surfaces, S stands alone—not grouped (notationally anyway) with the other orientable normal form surfaces. To alleviate this deficiency, we follow the convention of letting $0T$ denote the sphere S, and then the orientable normal form surfaces become those of the form nT ($n \geq 0$). Actually this convention is helpful in a "visual" sense: a torus, $1T$, can be continuously deformed so that it appears as a sphere with one "handle" attached. Likewise a two-holed torus, $2T$, can be viewed as a sphere with two handles attached, and so on. Since a sphere is just simply a sphere with no handles attached, $0T$ seems like a good name for it!

We will also be considering another collection of alternative nonorientable normal form surfaces, namely those of the form $P\#nT$ ($n \geq 0$) or $K\#nT$ ($n \geq 0$). Later the reader will be invited to show in an exercise that any nonorientable compact surface is homeomorphic to one of these, and we will discuss the relative "visual" merits of the two different families of nonorientable normal form surfaces.

3.2 THE ALGEBRA OF SURFACES

Before proving the classification theorem for compact surfaces, we will need to introduce some "algebraic" ideas. The main idea of this section is the representation of compact surfaces by *words*. A word is simply a list of symbols—say letters of the alphabet. We will see how words (or even portions of words) can be algebraically combined or modified in ways that suggest certain aspects of the surfaces they represent. Our goal will be to discover ways in which a word can be algebraically simplified without changing the surface represented. In the next section these procedures, called *circulation rules*, will be the main tools used in the proof of the classification theorem.

Sources variously attribute the first proof of a classification theorem for compact surfaces to either Dehn and Heegaard (1907) or H. R. Brahana (1922). Our approach, which uses words and circulation rules, can be directly traced to Brahana's 1922 paper "Systems of Circuits in Two-Dimensional Manifolds" [**Brahana**]. Since then, many authors have refined the role of—and simplified

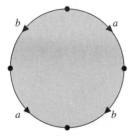

Figure 3.4 Four different words represent this plane model.

the notation for—words in the representation of surfaces; the terminology "circulation rules" appears in a 1960 paper by E. F. Whittlesey [**Whittlesey**]. In this book we will adopt the notation for words and the general form of the circulation rules introduced by P. A. Firby and C. F. Gardiner in their book *Surface Topology* [**Firby**].

The general idea of associating algebraic systems with objects in space is very important in the advanced study of topology. This section contains just the first of several such algebraic approaches presented in this book.

We have already noted that it is often easier to view a compact surface by considering one of its plane models rather than by looking directly at it as a set of points in space, and—thanks to the Triangulation Theorem—we can always find a plane model for any compact surface. But things can be simplified even further by letting compact surfaces be represented by certain types of algebraic expressions.

Let us illustrate what we want to do generally by looking at a familiar example. T—the torus—is represented by the simple plane model shown, once again, in Figure 3.4 as a circular disk. We will call a point on the boundary of a plane model where two of its edges meet a *vertex* of the plane model. (The plural of the word "vertex" is "vertices.") One vertex of this plane model is at the "12 o'clock" position on the disk. If we examine the edges as we move *clockwise* from 12 o'clock, we see edge a directed clockwise, edge b directed clockwise, edge a directed counterclockwise, and edge b directed counterclockwise. We are going to construct an algebraic expression, representing this plane model, by listing the edge labels in the clockwise order we saw them starting from the vertex at 12 o'clock. Further, if an edge is directed counterclockwise, we will include an exponent of –1 on its label in the expression. Thus, the expression we obtain is $aba^{-1}b^{-1}$. (Here a^{-1} is read "a inverse," as the reader might expect.) This expression is called a *word* for the plane model. By starting our clockwise examination of the edges at each of the other vertices (3 o'clock, 6 o'clock, or 9 o'clock) we can obtain three other words for the same plane model: $ba^{-1}b^{-1}a$, $a^{-1}b^{-1}ab$, and $b^{-1}aba^{-1}$. Clearly, given any one of these words, we could completely reconstruct the plane model it represents. And, in turn, since the plane model represents the compact surface T, each of these words can also be said to represent the *surface T.*

This procedure can be applied to obtain a word for any plane model.

> *Definition*
>
> A *word* representing a given plane model for a compact surface X is a list of the edge labels of the plane model read clockwise starting from some vertex and including an exponent of -1 on the edge label of any counterclockwise directed edge.
>
> Any such word is also said to *represent* the compact surface X.
>
> If M_1 and M_2 are two words representing homeomorphic compact surfaces, then M_1 and M_2 are said to be *homeomorphic* words, and we write $M_1 \sim M_2$.

Now, given a word for a plane model of a compact surface X, we can imagine traveling along the surface following the edges as specified by the word. We would begin at the point on the surface corresponding to the initial vertex of the first edge of the word, follow along that edge on the surface, then follow along the next edge, and so forth. Note that we will travel along each edge twice, and if an edge label x occurs in the word once as x and once as x^{-1} then our direction of travel will be reversed for the second pass along the edge. This gives us a useful visual device for interpreting the meaning of the algebraic symbol x^{-1}. Also note that we will end our trip along the surface at precisely the same point where we began. Thus, a word corresponds to a trip along a path on the surface, beginning and ending at some point (one of the vertices of the plane model) and following along edges of the plane model on the surface.

EXAMPLE 3.3

We have seen three different plane models for S, in Examples 2.8 and 2.9. These three plane models are represented by the words $abb^{-1}a^{-1}$, aa^{-1}, and $afg^{-1}e^{-1}b^{-1}bec^{-1}cgd^{-1}df^{-1}a^{-1}$, all of which are homeomorphic words since they all represent the compact surface S. The word aa^{-1} corresponds to taking a trip on the sphere: we start at the initial vertex of edge a, travel along a to its terminal vertex, and then retrace our steps back to the point where we started. The reader is encouraged to consider the trip on the sphere corresponding to each of the other words mentioned here.

When we look at the three words representing S in the previous example, one—aa^{-1}—appears to be the simplest. Indeed, except for a change in the actual letter used for the edge label, it is certainly the unique word of minimum length (using just two edges) which could represent S. We have also seen simple plane models, using a minimum number of edges for the compact surfaces T, K, and P. We will adopt the corresponding simple words as standard words for these four compact surfaces, and we will even refer to these words by the same capital letter, which denotes the surface represented.

> **Definition**
>
> S denotes the word aa^{-1}, T denotes the word $aba^{-1}b^{-1}$, K denotes the word $aba^{-1}b$, and P denotes the word aa.

We should note that words can similarly be used to represent sets that are not compact surfaces but can be constructed from plane models. For example, the cylinder is represented by the word $abcb^{-1}$, and the Möbius band is represented by the word $abcb$. Moreover, a reader who successfully solved Exercise S-1.3 will realize that $abcb \sim aab$.

From now on, an edge of a plane model tagged with the letter x and directed counterclockwise will be labeled x^{-1} and the arrow will usually not be shown.

• **Exercise 3.2**   For each of the following words, determine whether it represents an orientable compact surface, a nonorientable compact surface, or neither. In each case, explain your answer in terms of the edge labels of the word.

(a) $abca^{-1}c^{-1}b^{-1}$

(b) $abcda^{-1}c^{-1}d^{-1}$

(c) $abec^{-1}ba^{-1}cd^{-1}ed$

(d) $aa^{-1}bcab^{-1}c^{-1}$

(e) $abcfebacfe$

Since our classification theorem for compact surfaces will involve the notion of connected sums, we need to investigate the relationship between words and connected sums. The connected sum construction begins with the removal of an open disk from each of the compact surfaces being connected, and the disks can be removed from any location on the surface. Such a removal of an open disk can be accomplished on the plane model of a compact surface by simply removing an open disk from the interior of the plane model, and such a disk may be selected so that its boundary circle includes exactly one of the vertices of the plane model. Figure 3.5 shows such a situation on the standard plane model for T. The figure also shows how the "punctured" plane model for T can be spread out to become a plane model for the punctured torus. Note that the one vertex of the initial plane model that was also a point on the boundary circle of the removed disk has become *two* vertices of the final plane model. However, both of these vertices represent just one point on the actual punctured torus surface. The final plane model is a 5-gon, on which the single edge labeled x represents the boundary circle of the removed open disk. One resulting word that represents the punctured torus is therefore $aba^{-1}b^{-1}x$. Note that if we had selected a different vertex of the initial plane model to lie on the removed disk's boundary circle, we could have obtained some different words, such as $axba^{-1}b^{-1}$, $abxa^{-1}b^{-1}$ or $aba^{-1}xb^{-1}$.

Now suppose that we wish to form the connected sum of two compact surfaces, say for example two tori represented by words $aba^{-1}b^{-1}$ and $cdc^{-1}d^{-1}$. Given plane models corresponding to these words, we shall let the new inserted

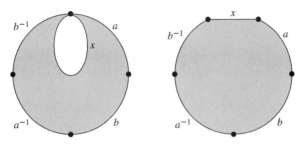

Figure 3.5 Obtaining a plane model for the punctured torus.

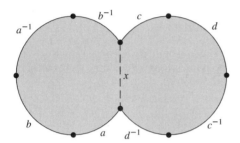

Figure 3.6 Obtaining a plane model for the connected sum of two tori.

edges introduced by puncturing each be denoted by the same edge label, say x. This is appropriate since in each case x represents the boundary circle of a removed disk, and when the connected sum is formed these boundary circles will be glued together as one curve on the connected sum. Figure 3.6 illustrates that the result of this gluing process for the two punctured plane models represented by the words $aba^{-1}b^{-1}x$ and $cdc^{-1}d^{-1}x$ is a plane model represented by the word $aba^{-1}b^{-1}cdc^{-1}d^{-1}$ in which the edge x has been "absorbed" into the interior. The resulting word, $aba^{-1}b^{-1}cdc^{-1}d^{-1}$, for the connected sum of two tori is very suggestive! By simply writing words for each of the two compact surfaces that are to be "summed" together, one followed by the other, we obtain a word that represents their connected sum. This process of combining two words into one word by writing one followed by the other is called concatenation.

Definition

The *connected sum* of two words M_1 and M_2 that represent compact surfaces is the word M_1M_2 obtained by writing M_1 followed by M_2. M_1M_2 is called the concatenation of M_1 and M_2.

The argument that the connected sum (concatenation) of two words for the torus yields a word representing the connected sum of two tori, is easily generalized to obtain the following result.

Theorem

If M_1 and M_2 are words that represent compact surfaces X_1 and X_2, then the word M_1M_2 represents the connected sum $X_1 \# X_2$ of the compact surfaces X_1 and X_2.

EXAMPLE 3.4

Now it's easy to write words representing even very involved compact surfaces if we know they are connected sums of simpler surfaces. For instance, a word that represents $T \# K \# P$, the connected sum of a torus, a Klein bottle, and a projective plane, is the word $TKP = aba^{-1}b^{-1}cdc^{-1}dee$.

• **Exercise 3.3**   Let X be the compact surface represented by the word $M = abc^{-1}ab^{-1}c$. Draw a sequence of pictures that show how the plane model of M and the standard plane models of the torus T and the projective plane P can be combined to obtain the word $abc^{-1}\,xyx^{-1}y^{-1}azzb^{-1}c$ representing $X \# T \# P$.

Our objective in the remainder of this section is to develop procedures, called *circulation rules*, for reducing a word for a compact surface to other words that are homeomorphic to the given word. Ultimately our goal will be to reduce a given word, in finitely many such steps, to a *normal form* word (i.e., a standard recognizable word for a normal form surface). Having accomplished this, we will be ready to prove the classification theorem for compact surfaces.

Before introducing the basic circulation rules, we need to discuss a little notation.

Definition

If a is an edge in a word, then $(a^{-1})^{-1}$ means a. If $A = a_1a_2...a_{k-1}a_k$ is a block of k consecutive edges in a word, then A^{-1} means $a_k^{-1}a_{k-1}^{-1}...a_2^{-1}a_1^{-1}$.

In light of our visual interpretation of a word providing a trip along the surface beginning and ending at some point, these symbolic notations are quite justified. The occurrence of $(a^{-1})^{-1}$ in a word would correspond to travel along the a edge in the direction opposite to that dictated by a^{-1}, that is, in the same direction as a itself would indicate. Also, A^{-1} would correspond to traveling on the surface along the edges of the block A, beginning with the last edge of A and traveling along each edge in the direction opposite to that indicated in the block A itself. The reader who has studied modern algebra will recall that the computation of the inverse of a (possibly noncommutative) product of invertible algebraic elements also follows this convention.

Note that a block, such as the block A in the definition above, of edges in a word for a compact surface could itself be a word for a compact surface, but this need not always be the case. Moreover, we might even allow a block referred to in a word to be empty.

We will develop five basic circulation rules for words. The first two rules are what we may call *positional circulation rules* in that they assert that changing the initial position of a plane model will not change the surface constructed when the gluings of the plane model are completed. The other three rules will be referred to as *cut-and-paste rules;* they propose that careful cutting apart of a plane model followed by a careful regluing will not change the surface produced.

The first positional circulation rule, which we will refer to as the *cycle rule,* simply asserts that a word for a compact surface is homeomorphic to any cyclic permutation of its edge labels. Essentially, the plane model represented by the word is rotated before the gluings are performed, and this will not change the surface produced. Note that a cyclic permutation of the edge labels of a word is equivalent to breaking the word into two blocks and exchanging the positions of those blocks—much like cutting a deck of cards simply cycles the cards around.

The second positional circulation rule is the *flip rule,* which says that any word for a compact surface is homeomorphic to its inverse. Imagine the plane model represented by a word as being a closed disk that can be lifted out of the plane, flipped over, and laid down again. The resulting plane model is now represented by the inverse of the original word, and certainly the gluings will produce the same surface that the original plane model produced. The positional circulation rules are summarized in the following theorem.

Theorem

Positional Circulation Rules for Words for Compact Surfaces

Let M be a word for a compact surface.
Cycle Rule: If $M = AB$, then $M \sim BA$.
Flip Rule: $M \sim M^{-1}$.

Note that either the cycle rule or the flip rule can be applied to a block within a word *if that block itself is a word for a compact surface.* In fact, this will be the case for all the circulation rules we will consider. To see this suppose that $M = ABC$ is a word for a compact surface, and that B is a block that is itself a word for a compact surface. By the cycle rule, $M \sim (CA)B$. Now the block CA must itself be a word for a compact surface too since CA contains an even number of edge labels, each occurring twice in CA. Thus, M represents the surface consisting of the connected sum of the compact surfaces represented by CA and B. So if D is a word for a compact surface and $B \sim D$, then $M \sim (CA)B \sim (CA)D \sim ADC$ where the last step follows from the cycle rule again.

Introduction of the "cut-and-paste" circulation rules makes us able to do some significant reduction of words for compact surfaces. The first such circulation rule is a word version of the fact that the connected sum of a given compact surface and a sphere is homeomorphic to the given surface. We call this circulation rule the *sphere rule*.

Theorem

The Cut-and-Paste Sphere Circulation Rule

If $M = Axx^{-1}B$ is a word for a compact surface where at least one of the blocks A or B is not empty, then AB is a word for a compact surface and $M \sim AB$.

The sphere rule allows us to delete any expression of the form xx^{-1} or $x^{-1}x$ from any location within a word for a compact surface, as long as some non-empty word remains. A simple algebraic justification of this rule involves applying the cycle rule so that the block xx^{-1} (or $x^{-1}x$) remains at the end of the word, which is then the concatenation of a word homeomorphic to AB with S and thus represents the connected sum of the compact surface represented by AB and the sphere. It follows at once that $M \sim AB$. Of course, expressions of the form xx^{-1} or $x^{-1}x$ could be inserted into words for compact surfaces as well, provided that the label x does not occur in the original word. The sense in which the sphere rule is a "cut-and-paste" rule is illustrated in Figure 3.7. There the plane model corresponding to the word $Axx^{-1}B$ is drawn as a square. The reader is invited to now construct and label a copy of this square and physically confirm that the x and x^{-1} edges can be folded to coincide. Gluing them—the reader may wish to use tape—produces a "cone" that is easily seen to be homeomorphic to a closed planar disk and, in fact, provides a plane model corresponding to the word AB.

We now have developed enough circulation rules to perform some interesting word reduction.

EXAMPLE 3.5

The word $M = afg^{-1}e^{-1}b^{-1}bec^{-1}cgd^{-1}df^{-1}a^{-1}$ was observed to represent the sphere in Example 3.3. We may now use circulation rules to formally reduce this word to the standard word S for the sphere, as follows.

$$M = afg^{-1}e^{-1}b^{-1}bec^{-1}cgd^{-1}df^{-1}a^{-1}$$
$$= afg^{-1}e^{-1}(b^{-1}b)e(c^{-1}c)g(d^{-1}d)f^{-1}a^{-1}$$
$$\sim afg^{-1}e^{-1}egf^{-1}a^{-1} \text{ (by sphere rule)}$$
$$\sim a^{-1}afg^{-1}e^{-1}egf^{-1} \text{ (by cycle rule)}$$
$$= (a^{-1}a)fg^{-1}(e^{-1}e)gf^{-1}$$
$$\sim fg^{-1}gf^{-1} \text{ (by sphere rule)}$$
$$= f(g^{-1}g)f^{-1}$$
$$\sim ff^{-1} = S \text{ (by sphere rule)}$$

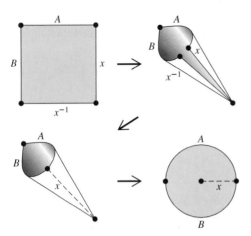

Figure 3.7 A "cut-and-paste" justification of the sphere rule.

The sphere rule justifies word reduction steps associated with *adjacent* edge labels x and x^{-1} within a word. We now consider words containing *nonadjacent* edge labels x and x^{-1}. In this case the plane model for such a word M can be drawn as a rectangle with the labels x and x^{-1} at two opposite ends, as shown in the first illustration in Figure 3.8. Note that if only the gluing of the x edges is performed, then those ends of the rectangle are joined and a cylinder is formed. The next "cut-and-paste" circulation rule we develop—the *cylinder rule*—asserts that a cyclic permutation of the edge labels of the intervening block between x and x^{-1} will not change the surface represented by the plane model. Now, since a cyclic permutation involves a partition of the block into two sub-blocks to be exchanged, let us assume the word considered is $M = AxBCx^{-1}D$ where BC is the block to be permuted into CB. As illustrated in Figure 3.8, cutting along a new edge y going from the vertex endpoint where A joins x to the vertex endpoint where B joins C and then gluing the x edges results in a plane model represented by the word $AyCBy^{-1}D$. This new plane model will produce the same surface as the original plane model when all gluings are performed. This result is summarized in the following theorem, in which we allow the symbolic substitution of the original edge label x for the new edge label y.

Theorem

The Cut-and-Paste Cylinder Circulation Rule

If M is a word for a compact surface and $M = AxBCx^{-1}D$, then $M \sim AxCBx^{-1}D$.

EXAMPLE 3.6

It is certainly the case that the word $M = abca^{-1}b^{-1}c^{-1}$ represents an orientable compact surface. To see what surface that is, we can use the circulation rules introduced thus far.

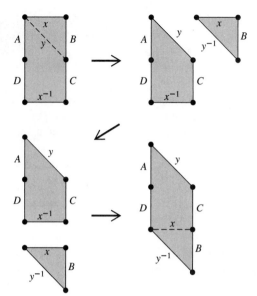

Figure 3.8 A "cut-and-paste" justification of the cylinder rule.

$$M = abca^{-1}b^{-1}c^{-1}$$
$$= a(bc)a^{-1}b^{-1}c^{-1}$$
$$\sim a(cb)a^{-1}b^{-1}c^{-1} \text{ (by cylinder rule)}$$
$$= ac(ba^{-1}b^{-1})c^{-1}$$
$$\sim ac(a^{-1}b^{-1}b)c^{-1} \text{ (by cylinder rule)}$$
$$= aca^{-1}(b^{-1}b)c^{-1}$$
$$\sim aca^{-1}c^{-1} \text{ (by sphere rule)}$$
$$= T$$

We have discovered that the word M represents the torus.

• **Exercise 3.4**   Use circulation rules introduced thus far to reduce each of the following words for orientable compact surfaces to a normal form word mT for some nonnegative integer m.

(a) $abcb^{-1}dc^{-1}d^{-1}a^{-1}$
(b) $aba^{-1}cdb^{-1}c^{-1}d^{-1}$

All four circulation rules introduced thus far can be applied to words that represent either orientable or nonorientable surfaces. We will need one more "cut-and-paste" rule, however, that applies only to reduction of words that represent nonorientable surfaces. Recall that a compact surface represented by a plane model is nonorientable if and only if there is at least one edge label such that the two edges with that label possess the same gluing direction, either both clockwise or both counterclockwise. Thus, a word for a nonorientable compact surface contains some edge labels, both occurrences of which have the same exponent—either both 1 or both −1.

The last circulation rule to be discussed concerns such words for nonorientable compact surfaces, that is, words of the form $M = AxBxC$ where A, B, and C are blocks of edge labels. The plane model for M can be drawn as a rectangle with the x edges at opposite ends, as shown in the first illustration of Figure 3.9. In gluing just the x edges together, the rectangle is given a half-twist and a Möbius band results. The cut-and-paste sequence of Figure 3.9 shows that if a cut is made along the indicated new edge y and the x edges are glued together, then a plane model with word $AyyB^{-1}C$ results that must produce the same surface. (A similar triangular plane model for a Möbius band was investigated in Exercise S-1.3.) A symbolic substitution of x for y yields our last circulation rule—the *Möbius band rule,* which is summarized in the following theorem.

Theorem

Cut-and-Paste Möbius Band Circulation Rule

If M is a word for a compact surface and $M = AxBxC$, then $M \sim AxxB^{-1}C$.

Read literally, this rule allows two similarly directed edge labels x in the word M to be made adjacent, with the intervening block B between them deleted and its inverse B^{-1} placed to the right of the now adjacent pair xx. The usefulness of such a reduction of a word lies in the fact that the adjacent pair xx now represents a projective plane whose connected sum with the rest of the word is represented by M.

EXAMPLE 3.7

The word $M = abca^{-1}b^{-1}c$ represents a nonorientable compact surface. To see exactly what surface that is, we apply circulation rules.

$$M = abca^{-1}b^{-1}c$$
$$= abc(a^{-1}b^{-1})c$$
$$\sim abcc(a^{-1}b^{-1})^{-1} \text{(by Möbius band rule)}$$
$$= abccba$$
$$\sim ccbaab \text{ (by cycle rule)}$$
$$= cc(baab)$$
$$\sim cc(aabb) \text{ (by cycle rule, applied to the block } baab\text{)}$$
$$= (cc)(aa)(bb)$$
$$= PPP = 3P$$

Actually, the compact surface represented by the word of the last example has several different interesting descriptions other than the connected sum of three projective planes. We will see this shortly.

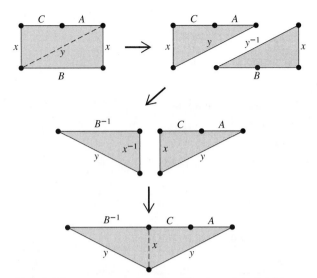

Figure 3.9 A "cut-and-paste" justification of the Möbius band rule.

• **Exercise 3.5**   Apply circulation rules to reduce each of the following words for non-orientable compact surfaces to a normal form word mP for some positive integer m.

(a) $abcba^{-1}c$
(b) $abec^{-1}ba^{-1}cd^{-1}ed$

EXAMPLE 3.8

The Klein bottle is homeomorphic to the connected sum of two projective planes, as the following computation shows.

$$
\begin{aligned}
K &= aba^{-1}b \\
  &= ab(a^{-1})b \\
  &\sim abb(a^{-1})^{-1} \text{ (by Möbius band rule)} \\
  &= abba \\
  &\sim aabb \text{ (by cycle rule)} \\
  &= (aa)(bb) \\
  &= PP = 2P
\end{aligned}
$$

The next example might seem rather bizarre—but it works! And it demonstrates the other descriptions of the compact surface of Example 3.7.

EXAMPLE 3.9

We claim that $T \# P$ is homeomorphic to $K \# P$, and in turn this is homeomorphic to $3P$. To obtain the result, apply the following sequence of circulation rules.

$$TP = (aba^{-1}b^{-1})(cc)$$
$$\sim a^{-1}b^{-1}(cca)b \text{ (by cycle rule)}$$
$$\sim a^{-1}b^{-1}cacb \text{ (by cylinder rule)}$$
$$= a^{-1}b^{-1}c(a)cb$$
$$\sim a^{-1}b^{-1}cca^{-1}b \text{ (by Möbius band rule)}$$
$$\sim (a^{-1}ba^{-1}b^{-1})(cc) \text{ (by cycle rule)}$$
$$\sim (bab^{-1}a)(cc) \text{ (by flip rule, applied to the block } a^{-1}ba^{-1}b^{-1})$$
$$= KP$$
$$\sim (PP)P \text{ (by the result of Example 3.8)}$$
$$= PPP = 3P$$

What is so amazing about this last example is that a Klein bottle, which is necessarily twisted up in 4-space, can apparently be untwisted *when a projective plane is attached to it via the operation of connected sum,* and that untwisting yields a *torus* with a projective plane attached!

3.3   THE CLASSIFICATION THEOREM: PRELIMINARY IDEAS

An early version of a classification theorem for compact surfaces was presented by Dehn and Heegaard in 1907. The proof that we will go through in this and the next section may be attributed to H. R. Brahana and dates back to the 1920s. The method of Brahana is based on the circulation rules that were introduced in the previous section and mathematical induction. Here is the statement of the theorem we will prove.

Theorem

The Classification Theorem for Compact Surfaces

(a) Any orientable compact surface is homeomorphic to mT for some nonnegative integer m, where $0T$ denotes the sphere.

(b) Any nonorientable compact surface is homeomorphic to mP for some positive integer m.

In order to prove this result, we will first call on the Triangulation theorem, which guarantees the existence of plane models for compact surfaces and, hence, allows us to concentrate on words for compact surfaces rather than the surfaces themselves. Thus, we will consider our proof to be complete when we have shown that any word for a compact surface can be reduced, using just finitely many applications of circulation rules, to a word for either mT ($m \geq 0$) in the orientable case or mP ($m \geq 1$) in the nonorientable case. We will refer to the surfaces mT ($m \geq 0$) and mP ($m \geq 1$) as *normal form surfaces* and will also refer to the words mT ($m \geq 0$) and mP ($m \geq 1$) as *normal form words.*

We will go through separate proofs of the orientable and nonorientable cases of the theorem, and the main structure of each of those proofs will be provided by mathematical induction. The *Principle of Mathematical Induction,* which we will now discuss briefly, is one of the most important, although quite elementary, techniques of mathematical proof. A reader who is already familiar with proof by induction may wish to skip this section and go on to the next section where we begin using induction to prove the classification theorem.

This technique of induction concerns proving that a statement involving an integer-valued variable n is true for all integers greater than or equal to some initial value n_0. For example, we might wish to prove that for all integers $n \geq 0$, any set with n elements has exactly 2^n subsets. (Recall that the empty set is always included as a subset of any set, as is the set itself.) We can check this assertion directly for small values of n: a set with 0 elements is the empty set and has exactly $1 = 2^0$ subset (itself), a set with 1 element has $2 = 2^1$ subsets (the empty set and the set itself), a set with 2 elements has $4 = 2^2$ subsets (the empty set, two one-element subsets, and the set itself), and so forth. But just checking the special cases where $n = 0$ or 1 or 2 or even the one thousand and one cases $n = 0, 1, 2, \ldots, 1000$ will not suffice to prove that the statement holds for all n.

The *Principle of Mathematical Induction,* which may be regarded as one of the axioms of arithmetic, asserts that the statement will be true for all $n \geq n_0$ if, in addition to checking its validity in the initial case $n = n_0$, we can also show that validity of the statement for an arbitrary, but fixed, integer k implies validity for the next integer $k + 1$. A somewhat contrived, but illuminating, device that illustrates this principle involves infinitely many dominoes lined up along the real line, one at each positive integer, as shown in Figure 3.10. Each domino is tall enough that it will knock over the one located one integer to the right if it falls that way. In this situation, if we consider the statement "The nth domino will be knocked over," then we of course expect all the dominoes to fall if we initiate the action by tipping the domino located at the integer 1 toward the positive direction.

So how about the statement "Any set with n elements has exactly 2^n subsets," which we would like to prove true for all integers $n \geq 0$? We have already verified this statement for $n = 0$, so let k be a fixed but arbitrary nonnegative integer. Now assume that we know that any set with k elements has exactly 2^k subsets. This assumption is called the *induction hypothesis.* (Note that in making the assumption of the induction hypothesis, we are not assuming the statement holds for all n—just for *one* fixed integer that we have called k.) We must prove, given the induction hypothesis, that any set with $k + 1$ elements has exactly 2^{k+1} subsets. So let a set S with $k + 1$ elements be given. Pick one element, say s, to be removed from S leaving a set T with k elements. By the induction

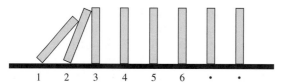

Figure 3.10 Mathematical induction with dominoes.

hypothesis, the set T has exactly 2^k subsets, each of which is also a subset of S. The other subsets of S must have s as an element and so must consist of a subset of T with the element s thrown in too. So there are precisely as many subsets of S containing the element s as there are subsets of T, accounting for a total of $2 \times 2^k = 2^{k+1}$ subsets of S, which is what we wanted to prove. Now the Principle of Mathematical Induction guarantees that, for all integers $n \geq 0$, any set with n elements has exactly 2^n subsets.

• **Exercise 3.6** Use the Principle of Mathematical Induction to prove that, for all positive integers n,

$$\sum_{i=1}^{n} i = \frac{n(n + 1)}{2}$$

There is an alternative version of proof by induction, called the *Principle of Complete Induction*, which is often quite useful. In this version of proof by induction, the induction hypothesis is taken to be the assumption that, for some fixed but arbitrary integer k, the statement being considered is valid for all integers less than k and greater than or equal to the initial case integer n_0. If, under this assumption, we can prove the statement's validity for the integer k, then the Principle of Complete Induction guarantees that the statement will hold for all $n \geq n_0$. The two versions of proof by induction can be shown to be equivalent, but we will simply accept them as two distinct, useful tools to be employed as needed in doing proofs.

An example in which the Principle of Complete Induction may be called upon in a natural way involves proving that every positive integer $n \geq 2$ is divisible by a prime number. In the initial case, $n = 2$, it is clear that the statement holds since 2 is a prime number and certainly divides itself. Now for some motivation for what to do next, let's consider the case of the integer 54, which is not prime. A nonprime integer greater than 1 must be composite, which means that it can be written as the product of two other integers strictly between 1 and itself, neither of which is necessarily prime. Now, of course, $54 = 2 \times 27$ and $54 = 3 \times 18$ and either of these factorizations shows that 54 is divisible by a prime number. But also $54 = 6 \times 9$ and, although neither 6 nor 9 is prime, each is itself divisible by a prime number. In a more general setting, if we consider a composite positive integer k and write $k = a \times b$, where a and b are integers strictly between 1 and k, then knowing that one of a or b is divisible by a prime number will force k to also be divisible by a prime number. The appropriate induction hypothesis here is that of the principle of complete induction, since the factors a and b may be much less than $k - 1$, as is so in the case where $k = 54$,

$a = 6$, and $b = 9$. So the remaining part of the induction proof would go like this: given an arbitrary, but fixed, integer $k > 2$, assume that every positive integer less than k and greater than or equal to 2 is divisible by a prime number. Now either k is prime or not. If k is prime, then certainly k is divisible by a prime number, namely k itself. If k is not prime, then it is composite and we can write $k = a \times b$, where a and b are integers strictly between 1 and k. By the induction hypothesis, either of a or b is divisible by a prime number, and therefore so is k. The Principle of Complete Induction now applies to assure us that any integer $n \geq 2$ is divisible by a prime number.

- **Exercise 3.7**   The Fibonacci numbers F_n are defined recursively for $n \geq 1$ by $F_1 = 1$, $F_2 = 1$, and $F_{n+2} = F_{n+1} + F_n$ for $n \geq 1$. Use the Principle of Complete Induction to prove that every Fibonacci number is an integer.

3.4   THE PROOF OF THE CLASSIFICATION THEOREM

We will prove the two parts of the Classification Theorem stated in the last section separately, first considering words for orientable compact surfaces and then proceeding to consideration of words for nonorientable compact surfaces. The proof of each part will involve an application of the Principle of Complete Induction. By the *length* of a word for a compact surface, we will mean the number of letters in it, which also equals the number of edges (or vertices) of its plane model.

We will first prove that if M is a word for an orientable compact surface and the length of M is $2n$, then $M \sim mT$ for some $m \geq 0$ satisfying $4m \leq 2n$. This last condition will be an automatic consequence of the fact that our proof will involve *reducing* the length ($2n$) of the original word M to obtain the word mT (which has length $4m$). We will apply the Principle of Complete Induction to do this, using n as the "induction variable." First consider the case in which $n = 1$. In this case, either $M = aa^{-1}$ or $M = a^{-1}a$, and from either of these we conclude that $M \sim S = 0T$.

Now let $k > 1$ be a fixed integer and assume that any word of length l less than $2k$ for an orientable compact surface is homeomorphic to mT for some m with $0 \leq m$ and $4m \leq l$. Let M be a word for an orientable compact surface and suppose that the length of M is $2k$. A pair of two edges x and y of M are called *alternating* if the references to x and y alternate in the word M so that, after a cyclic permutation as allowed by the cycle rule, $M \sim xAyBx^{-1}Cy^{-1}D$, where A, B, C, and D are blocks of edges of M, some possibly empty blocks. Consider the case where M has no alternating pairs of edges. In this case, let a be the first edge of the word M. Now $M = aAa^{-1}B$, where A and B are blocks of edges of M and at least one of A or B is nonempty since the length of M is greater than 2. Note that if either A or B is a nonempty block, then that block must itself be a word for an orientable compact surface. This is true since no edge y of M can appear in both A and B, for otherwise a and y would be alternating edges of M. Now if A is empty, then $M = aa^{-1}B \sim B$ (using the sphere rule). Similarly if B is empty, then $M = aAa^{-1} \sim a^{-1}aA \sim A$ (using the cycle and sphere rules). Further, if both A and B are nonempty, then $M = aAa^{-1}B = (aAa^{-1})B \sim (a^{-1}aA)B \sim AB$

(using the cycle and sphere rules applied just to the indicated grouped blocks, which are themselves words for compact surfaces). In any case, we have shown that M is homeomorphic to a word of length less than $2k$ for an orientable compact surface that, by the induction hypothesis, is in turn homeomorphic to mT for some $m \geq 0$ with $4m \leq 2k - 2 < 2k$.

Now suppose that M has an alternating pair x and y of edges, so that $M \sim xAyBx^{-1}Cy^{-1}D$. Then

$$M \sim xAyBx^{-1}Cy^{-1}D$$

$$= xAy(Bx^{-1}C)y^{-1}D$$

$$\sim xAy(CBx^{-1})y^{-1}D \text{ (by cylinder rule)}$$

$$\sim y^{-1}DxAyCBx^{-1} \text{ (by cycle rule)}$$

$$= y^{-1}(DxA)yCBx^{-1}$$

$$\sim y^{-1}(xAD)yCBx^{-1} \text{ (by cylinder rule)}$$

$$= y^{-1}x(ADyCB)x^{-1}$$

$$\sim y^{-1}x(CBADy)x^{-1} \text{ (by cylinder rule)}$$

$$\sim (yx^{-1}y^{-1}x)(CBAD) \text{ (by cycle rule)}$$

$$\sim (xyx^{-1}y^{-1})(CBAD) \text{ (by cycle rule)}$$

$$= T(CBAD).$$

Now clearly $CBAD$ is a word for an orientable compact surface and the length $(2k-4)$ of $CBAD$ is less than $2k$. By the induction hypothesis, $CBAD \sim mT$ for some $m \geq 0$ with $4m \leq 2k-4$. Consequently $M \sim TmT = (m + 1)T$, and $4(m + 1) = 4m + 4 \leq (2k-4) + 4 = 2k$.

Our work in the two cases considered allows us to apply the Principle of Complete Induction to conclude that any word of length l for an orientable compact surface is homeomorphic to mT for some nonnegative integer $m \geq 0$ satisfying $l \geq 4m$. This completes the proof of the orientable part of the classification theorem.

We now need to prove that if M is a word for a nonorientable compact surface and the length of M is $2n$, then M is homeomorphic to a word of the form mP for some $m \geq 1$. Again we will use the Principle of Complete Induction. First considering the case where $n = 1$, we immediately see that $M = aa = P = 1P$. So let $k > 1$ be fixed, and assume that any word of length less than $2k$ for a nonorientable compact surface is homeomorphic to a word of the form mP for some $m \geq 1$. Let M be a word for a nonorientable compact surface and suppose the length of M is $2k$. With a simple cyclic permutation of the edges of M (as allowed by the cycle rule), we can find an edge a of M such that $M \sim aAaB$, where A and B are blocks of edges of M. Then, by the Möbius band rule, $M \sim aaA^{-1}B = (aa)(A^{-1}B) = P(A^{-1}B)$, and $A^{-1}B$ is certainly a word for a compact surface of length less than $2k$.

Now $A^{-1}B$ could either represent an orientable compact surface or a nonorientable compact surface. In the nonorientable case, by the induction hypothesis, $A^{-1}B \sim mP$ for some positive integer m, and it follows that $M \sim P(A^{-1}B) \sim PmP = (m + 1)P$. On the other hand, if $A^{-1}B$ represents an orientable compact surface, then we may apply the (already proven) classification of orientable compact surfaces to conclude that $A^{-1}B \sim rT$ for some $r \geq 0$ and satisfying $4r < 2k$. If $r = 0$, then $M \sim P(0T) = PS \sim P = 1P$. If $r > 0$, then

$$M \sim P(A^{-1}B)$$

$$\sim P(rT)$$

$$= (PT)((r - 1)T)$$

$$= (PK)((r - 1)T) \text{ (by the results of Example 3.9)}$$

$$\sim P(K(r - 1)T).$$

Now $K(r-1)T$ represents a nonorientable compact surface and is a word of length $4r < 2k$. So the induction hypothesis guarantees that $K(r-1)T \sim mP$ for some $m \geq 1$, and in turn $M \sim PmP = (m + 1)P$.

So the Principle of Complete Induction allows the conclusion that any word for a nonorientable compact surface is homeomorphic to a word of the form mP for some positive integer m. Now the proof of both parts of the classification theorem is complete.

The remarkable simplicity of the normal form compact surfaces mT ($m \geq 0$) and mP ($m \geq 1$) cannot be denied. And the result provided by the Classification Theorem, which declares that any compact surface is homeomorphic to some normal form compact surface, is indeed quite elegant. The fact that nonorientable compact surfaces cannot exist without self-intersections in 3-space is somewhat of an obstacle when we try to visualize such surfaces, and being able to predict that any such surface may be viewed as a connected sum of projective planes provides some consolation. In fact, just as we have thought of the orientable normal form surface mT as a sphere with m nice "handles" attached to it, we can also think of the nonorientable normal form surface mP as a sphere with m "cross-caps" attached to it. Here, the word "cross-cap" refers to a punctured projective plane (i.e., Möbius band) whose boundary circle can be attached to the boundary circle of a removed open disk on the sphere.

However, a normal form compact surface of the form mP "goes into 4-space" in m different places on the surface, and if m is very large, then this does not help our intuition much when we are thinking about the surface. Further relief in this respect may be obtained by considering an alternative set of normal form surfaces for the nonorientable compact surfaces. The proof of the following result is left to the reader to complete as a supplementary exercise at the end of this chapter. Some indication that this result holds actually arose in the proof of the nonorientable case of the Classification Theorem.

> *Theorem*
>
> Any nonorientable compact surface is homeomorphic to an *alternative nonorientable normal form surface* of the form *K#mT*, for some nonnegative integer *m*, or *P#mT*, for some nonnegative integer *m*.

We have seen simple ways of visualizing a single Klein bottle *K* and a single projective plane *P*, and we can understand easily how the apparent self-intersection in each can be avoided in 4-space. Any of these alternative normal form surfaces, being the connected sum of either *K* or *P* with a multi-holed torus, "goes into 4-space" at exactly one location on the surface and does so in a simple way that we can predict. Further, since *K* is homeomorphic to 2*P*, the nonorientable compact surfaces can now be visualized as multi-holed (allowing zero-holed) tori with either one or two cross-caps attached.

• **Exercise 3.8**   For each of the following words for nonorientable compact surfaces, determine an alternative normal form word of the form *KmT* or *PmT* to which the given word is homeomorphic.

(**a**) *abcabc*

(**b**) 4*P*

3.5   CHAPTER SUMMARY

In this chapter we have seen how any compact surface can be constructed as a connected sum of some very simple compact surfaces. In particular, any orientable compact surface has been shown to be homeomorphic to a sphere or a multi-holed torus, and we have come to know these surfaces as the orientable normal form surfaces. Similarly, we have seen that any nonorientable compact surface is homeomorphic to a nonorientable normal form surface consisting of a connected sum of finitely many projective planes or to an alternative nonorientable normal form surface consisting of a connected sum of either a projective plane or a Klein bottle with either a sphere or a multi-holed torus. The method developed to prove the classification theorem for compact surfaces is a process in which topological notions associated with objects in space have been expressed in terms of algebraic notions associated with words representing those objects, and the link between the topology and the algebra was the idea of plane models. The particular tools used in the proof included circulation rules for words for compact surfaces and mathematical induction.

There is one further question to ask about the result of the classification theorem. This theorem guarantees that a given compact surface is homeomorphic to some normal form compact surface. But could it possibly be homeomorphic to several different normal form surfaces? This question is equivalent to the question that asks whether any different normal form surfaces could be homeomorphic to each other. Certainly an orientable normal form surface will not be homeomorphic to any nonorientable normal form surface. But could 5*T* somehow turn out to be homeomorphic to $7T$? Or could $67P$ be homeomorphic to $45P$? In the next chapter we will learn about a number—the *Euler characteristic*—associated with a compact surface, that will help us answer this question. And the Euler characteristic will also be a primary tool in our investigation of tilings of surfaces by polygons—a problem that has been of interest to mathematicians from ancient times to the present.

3.6   SUPPLEMENTARY EXERCISES

Exercise S-3.1   In this chapter it was proven that K is homeomorphic to $2P$ and that $K\#P$ is homeomorphic to $T\#P$. Using just these two facts and commutativity and associativity of the connected sum operation $\#$, find the normal form surface of the form mP that is homeomorphic to $T\#(2K)\#T\#P\#K\#P$.

Exercise S-3.2   Use circulation rules to reduce each of the following words for compact surfaces to a normal form word mT or mP.
(a) $ab^{-1}c^{-1}a^{-1}cb$
(b) $abc^{-1}bca$
(c) $ab^{-1}cedefa^{-1}bc^{-1}d^{-1}f$

Exercise S-3.3   For each positive integer n, let M_n be the word defined by $M_n = a_1a_2$ $\ldots a_na_1^{-1}a_2^{-1}\ldots a_n^{-1}$. It is clear that $M_1 \sim S$ and $M_2 \sim T$, and it was shown in Example 3.6 that $M_3 \sim T$ also.
(a) Show that $M_4 \sim 2T$.
(b) Show that, for a fixed but arbitrary positive integer $n \geq 1$, $M_{n+2} \sim TM_n$. (Note that this is *not* asking for an induction proof.)
(c) Use the Principle of Mathematical Induction to show that for all integers $n \geq 1$, $M_{2n} \sim nT$ and $M_{2n+1} \sim nT$.

Exercise S-3.4   Use circulation rules (and associativity and/or commutativity of connected sums) along with the Principle of Mathematical Induction to show that for all integers $m \geq 0$, $KmT \sim 2(m+1)P$ and $PmT \sim (2m+1)P$. Then explain why any word representing a nonorientable compact surface is homeomorphic to an alternative normal form word.

Chapter 4

Putting More Structure on Surfaces

In this chapter we will consider some important questions concerning how a surface can be "tiled by" (or constructed from) certain types of its subsets. Some of these questions—and some of their answers—have been around since ancient times, while some have been answered only very recently. The ideas of this chapter have led to the development of whole new fields of mathematical study and have found some interesting applications.

The first two sections of the chapter are devoted to investigating tilings on compact surfaces. The third section concerns the notion of coloring maps on compact surfaces. Throughout all these sections, a very special invariant of compact surfaces called the *Euler characteristic* will prove to be an invaluable tool.

4.1 TILINGS ON SURFACES AND THE EULER CHARACTERISTIC

When we first discussed the sphere as a surface in Chapter 1, we noted that it is topologically equivalent to the surface of a cube, which is naturally viewed as six squares whose edges are glued together appropriately. Also in Chapter 1, we considered the surface of a donut—the torus—as the surface of a simple picture frame (see Figure 1.5), which we viewed as a collection of sixteen four-sided faces with edges glued together in a special way. In each of these cases, the object of our attention was simply a compact surface (the sphere or the torus), but some extra "structure" was provided by the way we viewed that surface. That structure consisted of some specified subsets of the surface (faces, edges, and vertices) combined in a special way so as to cover the surface. Such a covering of a compact surface is called a *tiling* on the surface.

Some more formal terminology is needed to make this idea precise. By a *closed cell* we will mean any set of points that is homeomorphic to a closed planar disk, and we will use *open cell* to mean any set of points homeomorphic to an open planar disk. A closed cell becomes a *polygon* when finitely many, say a where $a \geq 1$, points on its boundary are specified as *vertices* of the polygon, and the a arcs of the boundary joining consecutive vertices are then called *edges* of

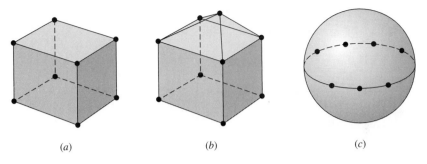

Figure 4.1 Three tilings on the sphere.

the polygon. A polygon with a vertices (and, thus, a edges) is called an *a-gon*— terminology we have already used in other contexts. Thus, a triangle is a 3-gon, a rectangle is a 4-gon, a pentagon is a 5-gon, and so on.

Definition

A *tiling* on a compact surface X is an arrangement of finitely many polygons on X satisfying

(a) the polygons cover X, and
(b) if polygons meet, they do so either at vertices or along complete edges.

The polygons of a tiling are called the *faces* of the tiling, and the vertices and edges of the polygons are called the *vertices* and *edges* of the tiling.

EXAMPLE 4.1

Figure 4.1 shows illustrations of three tilings on the sphere. Tiling (a) is the sphere supplied with the extra structure of a cube's surface: 6 square faces accounting for 8 vertices and 12 edges. The faces of tiling (b) consist of 5 squares and 4 triangles, and this tiling has totals of 9 vertices and 16 edges. Tiling (c) has just two 7-gon faces and has 7 vertices and 7 edges.

We have already learned that a plane model of a compact surface can simplify visualizing the surface, and a sketch of a tiling's vertices and edges on a plane model is often a very convenient way to illustrate the tiling.

EXAMPLE 4.2

Figure 4.2 shows a tiling on the usual (rectangular) plane model of the Klein bottle, K, consisting of 8 rectangles and 2 hexagons and containing 12 vertices and 22 edges. Note that the four corners of the plane model (which represent just *one* point on the actual

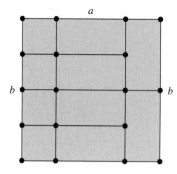

Figure 4.2 A tiling on the Klein bottle.

surface) also represent one of the vertices of the tiling, and also the four edges of the plane model are covered by edges of the tiling.

In the examples above, we carefully counted the numbers of vertices, edges, and faces of each tiling. These numbers are important enough to merit recognition by special notation.

Definition

For a tiling on a compact surface

V denotes the number of vertices of the tiling,
E denotes the number of edges of the tiling, and
F denotes the number of faces of the tiling.

EXAMPLE 4.3

Refer to Example 4.1 where three tilings on the sphere were introduced. For tiling (a), $V = 8$, $E = 12$, and $F = 6$. For tiling (b), $V = 9$, $E = 16$, and $F = 9$. And for tiling (c), $V = 7$, $E = 7$, and $F = 2$.

• **Exercise 4.1**   A tiling on the projective plane by 3-gon and 4-gon faces is shown on the plane model of P in Figure 4.3. Note that the vertices and edges of the tiling are shown as heavier lines and dots than are the vertices and edges of the plane model itself. For this tiling, determine the numbers of 3-gons and 4-gons of the tiling, and also determine V, E, and F.

It is of interest to compute the value of a special combination of V, E, and F for each of the tilings on the sphere in Example 4.3. In each case, as the reader can easily check, $V - E + F$ yields the same number, namely 2. However, for the tiling on the Klein bottle in Example 4.2, where we had $V = 12$, $E = 22$, and $F = 10$, it

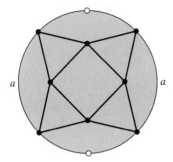

Figure 4.3 The tiling on the projective plane for Exercise 4.1.

turns out that $V - E + F = 0$. It is perhaps not surprising that $V - E + F$ did not yield 2 this time since the tiled surface was not the sphere, and the reader might expect that other tilings on K might also satisfy $V - E + F = 0$.

EXAMPLE 4.4

Consider the tiling on K consisting of 4 rectangles as shown on the plane model in Figure 4.4. Being careful again about observing which vertices and edges appear more than once on the boundary of the plane model, we see that $V = 4$, $E = 8$, and $F = 4$. Thus, $V - E + F = 0$ for this tiling on K, in agreement with the result for the tiling of Example 4.2.

• **Exercise 4.2**   Compute $V - E + F$ for the tiling on the projective plane which was considered in Exercise 4.1. What would you expect the value of $V - E + F$ to be for any other tiling on P?

These examples and exercises suggest a general result, which we will now state as a theorem. We will discuss the proof of this theorem later in this section, following an investigation of a special case.

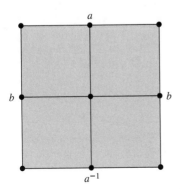

Figure 4.4 Another tiling on the Klein bottle.

Theorem

Basic Tiling Conjecture for Compact Surfaces

Let X be a compact surface. Then there is an integer, denoted by $\chi(X)$ and called the *Euler characteristic* of X, such that for any tiling on X, $V - E + F = \chi(X)$.

The definition of tiling does not provide any restrictions on how often the polygonal faces may meet on a surface or, for that matter, whether a face may meet itself. For example, a $2n$-gon plane model for a compact surface X provides a tiling on X consisting of just one $2n$-gon face, which meets itself many times on the surface X when the gluings of the plane model's edges are made. Now the number E of edges of this tiling, as it appears on the surface X, is n since the edges of the plane model are identified in pairs. Also, when the gluings are made, some of the $2n$ vertices of the plane model may get glued together, thus possibly producing fewer than $2n$ distinct points on the surface. If we let v denote the number of these distinct points, then for this tiling we will have $V = v$ and, hence, $V - E + F = v - n + 1$, which is completely determined by the plane model, or even by any word representing the plane model. This leads us to define the *Euler characteristic* for a word for a compact surface.

Definition

Let M be a word of length $2n$ for a compact surface, and let v denote the number of distinct points on the compact surface corresponding to the $2n$ vertices of the plane model associated with M. Then the *Euler characteristic* of M is the number $\chi(M) = v - n + 1$.

EXAMPLE 4.5

For the word $S = aa^{-1}$ of length 2, $n = 1$ and, since the initial and terminal ends of the edge a correspond to different points on the surface of the sphere, $v = 2$. So $\chi(S) = v - n + 1 = 2 - 1 + 1 = 2$. The word $M = afg^{-1}e^{-1}b^{-1}bec^{-1}cgd^{-1}df^{-1}a^{-1}$ is of length 14 so that for this word $n = 7$. Recalling from Example 2.9 that the vertices of the plane model of M correspond to all 8 corners of a cube, we see that $v = 8$ and so $\chi(M) = v - n + 1 = 8 - 7 + 1 = 2$.

Of course, in this example the words M and S both represent the sphere, so the value of 2 here should not be surprising, since it is the value of $V - E + F$ we have seen previously for tilings on the sphere.

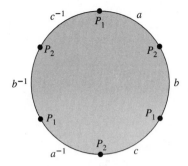

Figure 4.5 The vertices of this plane model represent just two points.

EXAMPLE 4.6

In Example 3.6, we showed that $M = abca^{-1}b^{-1}c^{-1}$ is homeomorphic to the word $T = aba^{-1}b^{-1}$, so that both represent the torus. Let us compute the Euler characteristic for M and T.

One can easily check, on the plane model of M shown in Figure 4.5, that $v = 2$. Since M is of length 6, $n = 3$. So $\chi(M) = v - n + 1 = 2 - 3 + 1 = 0$.

Now T has length 4, so $n = 2$. And on the plane model for T (which is the standard plane model for the torus) all 4 vertices of the plane model are glued together as one point. So $v = 1$ and $\chi(T) = v - n + 1 = 1 - 2 + 1 = 0$. Thus, $\chi(M)$ and $\chi(T)$ are equal, and their value is 0.

• **Exercise 4.3**   For each of the given words for compact surfaces, compute the Euler characteristic of the word directly from the word and its plane model.

(a) $ab^{-1}c^{-1}dd^{-1}cba^{-1}$

(b) $abc^{-1}adb^{-1}cd$

(c) $abccb^{-1}a$

Based on Example 4.6, and the tiling conjecture, we would expect the Euler characteristic of any word representing the torus to be 0 and would also expect that for any tiling of the torus we would have $V - E + F = 0$.

It is now appropriate to consider the proof of the special case of the tiling conjecture *for words* (i.e., for tilings induced by plane models). Recall that, by definition, two words that represent the "same" compact surface (i.e., homeomorphic surfaces) are said to be *homeomorphic* words. Now a useful way of converting a first word into another word—homeomorphic to the first—is to use finitely many applications of the circulation rules (along with commutativity and associativity of connected sum). A question that lingers here asks whether, given two homeomorphic words M_1 and M_2, finitely many applications of the circulation rules can be used to convert M_1 into M_2. The answer to this question is in the affirmative, although the proof (and generalization) of this result will not be given here. This result follows from a theorem called the *Hauptvermutung* (a German word meaning "principal conjecture"), which was first stated in

1907 by Steinitz and proven in 1923 by Kerekjarto. We shall borrow the name *Hauptvermutung* to refer to the special result we will need.

Theorem

The Hauptvermutung

Let M_1 and M_2 be words for compact surfaces and assume that $M_1 \sim M_2$ (i.e., M_1 and M_2 represent homeomorphic compact surfaces). Then M_1 can be converted into M_2 using finitely many applications of the circulation rules and commutativity and associativity of connected sums.

Now, getting back to the notion of Euler characteristic for words, let us start with two homeomorphic words M_1 and M_2. Then we know that we can use circulation rules to convert M_1 into M_2 in finitely many steps, as guaranteed by the Hauptvermutung, so we will be sure that M_1 and M_2 have the same Euler characteristic if none of these steps ever changes an Euler characteristic. If we concentrate on the numbers v and n, a little thought will convince us that the only sort of step that would change either is the sphere circulation rule. But any one application of the sphere rule causes v and n to *both* increase by 1 or to *both* decrease by 1. In either case, the Euler characteristic, given by $v - n + 1$ remains the same. Let us summarize this useful result as another theorem.

Theorem

If M_1 and M_2 are two words for compact surfaces and $M_1 \sim M_2$, then $\chi(M_1) = \chi(M_2)$. That is, the Euler characteristic of a word, as well as $V - E + F$ for the tiling of its plane model, depends only on the compact surface that the word represents.

This result concerning tilings induced by plane models is a first step in one approach to proving the general tiling conjecture. We will not go through the technical steps of this proof but instead just provide an outline. Starting with any tiling on a compact surface X, suppose that we can use the tiling to produce a plane model for X by cutting along some of the edges of the tiling as needed. (This may require some preliminary modification of the pattern in certain extreme cases. But, for our purposes here, we will assume this can be accomplished.) The plane model so produced will have its $2n$ boundary edges representing n of the E edges of the tiling and its $2n$ vertices will represent some number, v, of the tiling's V vertices. Now, if we can carefully remove—step by step—the edges and vertices of the tiling that appear on the *interior* of the plane model without changing the quantity $V - E + F$ in any step, then we will be

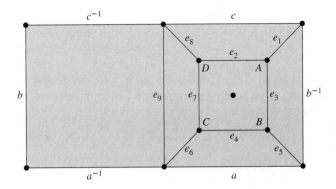

Figure 4.6 Removing interior edges and vertices does not alter $V - E + F$.

assured that the value of $V - E + F$ for the original tiling equals $V - E + F$ for the plane model itself. And that value is $v - n + 1$. Thus, the Euler characteristic depends only on the surface that is tiled, not on the specific tiling for which $V - E + F$ is computed.

Let us appeal to an example to illustrate this process.

EXAMPLE 4.7

In Figure 4.6 the tiling on the sphere given by the cube surface is considered. Cutting along three of the edges of its upper face and spreading out the resulting set produces the 6-sided plane model shown. There are 9 edges and 4 vertices of the tiling that appear in the interior of that plane model. Removal of edge e_1 leaves one face where two had been previously; so E and F are both reduced by 1 and $V - E + F$ is unchanged. Next, removal of edges e_2 and e_3 along with vertex A leaves one face where there were two previously, thus reducing V by 1, E by 2, and F by 1. The net change in $V - E + F$ is $-1 - (-2) - 1 = 0$. The remaining steps involve removing, in order, the following groups of edges and vertices: $\{e_4, e_5, B\}$, $\{e_6, e_7, e_8, C, D\}$, and $\{e_9\}$. The reader can easily check that these steps produce no change in $V - E + F$.

It is interesting to note that, for tilings on the sphere, the result $V - E + F = 2$ has been known since ancient times but was proven formally by Euler just a few centuries ago. The fact that $V - E + F$ is an invariant for patterns on the sphere led to the more modern investigations of the Euler characteristic for general compact surfaces.

A few final comments should be made about the tiling conjecture and its proof. The condition in the definition of a tiling requiring the faces to be polygons—that the interiors of the faces be open cells—is needed here. Thus, for example, if an arrangement of vertices and edges on the sphere divides the sphere into closed regions, at least one of which is not a polygon, then it is possible that $V - E + F$ will not equal 2. Also, we have used the Hauptvermutung in our proof of the special case of the tiling conjecture for words primarily to emphasize the nontrivial nature of the fact that homeomorphic words can be converted one into the other by way of circulation rules. In the more advanced study of topology, the general Hauptvermutung plays an extremely important role. However, the reader should be aware that in the specific case of tilings on compact connected 2-dimensional manifolds the tiling conjecture can be proven using standard techniques of general topology without assuming the quite powerful result asserted by the general Hauptvermutung. These techniques, which are somewhat beyond the intended scope of this textbook, might be of interest to the reader for futher study.

The Euler characteristic of a compact surface can be determined from a word representing the surface, and we know—from results of Chapter 3—that all compact surfaces are represented by normal form words. Thus, we can now easily compute the Euler characteristic of any compact surface from a normal form word representing it.

EXAMPLE 4.8

We can compute the Euler characteristic of each of the elementary compact surfaces S, T, K, and P by computing the Euler characteristics of their standard words $S = aa^{-1}$, $T = aba^{-1}b^{-1}$, $K = aba^{-1}b$, and $P = aa$. We have already done so for S and T in previous examples, and so we obtain $\chi(S) = 2$ and $\chi(T) = 0$. It is left to the reader to confirm that $\chi(K) = \chi(aba^{-1}b) = 0$ and that $\chi(P) = \chi(aa) = 1$.

The reader might have been surprised by something in the last example. The Euler characteristics for the torus and the Klein bottle are both 0, even though the torus and the Klein bottle are not homeomorphic! So it is clear that, although two homeomorphic compact surfaces must have the same Euler characteristic, sometimes two nonhomeomorphic compact surfaces may also have the same Euler characteristic.

EXAMPLE 4.9

Let M be any normal form word other than S, that is, one of the form mT ($m \geq 1$) or mP ($m \geq 1$). For any of these normal form words all vertices of its plane model are glued together as one point on the surface represented, so $v = 1$. Now mT is of length $4m$, so for mT, $n = 2m$ and $\chi(mT) = v - n + 1 = 1 - 2m + 1 = 2 - 2m$. (Note that if $m = 0$ then $2 - 2m = 2$, so this formula works for $0T = S$ as well.)

Also, the length of mP is $2m$. So for mP, $n = m$ and $\chi(mP) = v - n + 1 = 1 - m + 1 = 2 - m$.

EXAMPLE 4.10

This time, let M be any alternative normal form word for a nonorientable compact surface other than P or K, that is, KmT or PmT where $m \geq 1$. For all such words, again $v = 1$, the length of KmT is $4(m + 1)$, and the length of PmT is $2 + 4m$. Thus,

$\chi(KmT) = v - n + 1 = 1 - 2(m + 1) + 1 = -2m$, and $\chi(PmT) = v - n + 1$
$= 1 - (1 + 2m) + 1 = 1 - 2m$. (Note again that these results also hold if $m = 0$.)

In the last few examples we have proven a theorem that gives a complete summary of the Euler characteristics of all compact surfaces.

Theorem

If $m \geq 0$, then $\chi(mT) = 2 - 2m$.
If $m \geq 1$, then $\chi(mP) = 2 - m$.
If $m \geq 0$, then $\chi(K \# mT) = -2m$.
If $m \geq 0$, then $\chi(P \# mT) = 1 - 2m$.

• **Exercise 4.4**   If there exists a tiling on the four-holed torus $4T$ with exactly 52 vertices and exactly 58 faces, then how many edges must the tiling have?

This theorem tells us everything there is to know about the numbers that arise as Euler characteristics of compact surfaces. For example, the Euler characteristics of orientable compact surfaces are the even integers less than or equal to 2. Also, no two orientable normal form compact surfaces $m_1 T$ and $m_2 T$, where $m_1 \neq m_2$, can be homeomorphic. For if they were, then $2 - 2m_1 = \chi(m_1 T)$ $= \chi(m_2 T) = 2 - 2m_2$ would imply that $m_1 = m_2$.

The theorem also tells us that any integer less than or equal to 1 is the Euler characteristic of a nonorientable compact surface. An argument essentially identical to that in the last paragraph shows that no two nonorientable normal form surfaces $m_1 P$ and $m_2 P$, where $m_1 \neq m_2$, can be homeomorphic. It is left as an exercise (see Exercise S–4.3 in the Supplementary Exercises at the end of this chapter) to show directly that no two different alternative nonorientable normal form surfaces can be homeomorphic.

Now if two compact surfaces are homeomorphic, then they both must be orientable or they both must be nonorientable. Also, they must have the same Euler characteristic. Conversely, if two compact surfaces with the same Euler characteristic are both orientable or both nonorientable, then they must be homeomorphic. This can be proved in a similar fashion for the orientable or nonorientable case. For the orientable case, suppose that X and Y are orientable compact surfaces and that $\chi(X) = \chi(Y)$. By the classification theorem, X is homeomorphic to $m_1 T$ and Y is homeomorphic to $m_2 T$ for some nonnegative integers m_1 and m_2. Now $2 - 2m_1 = \chi(m_1 T) = \chi(X) = \chi(Y) = \chi(m_2 T) = 2 - 2m_2$, so that we may conclude that $m_1 = m_2$. (The similar proof in the nonorientable case involves connected sums of projective planes, of course.) Thus, Euler characteristic and orientability determine compact surfaces completely.

> ***Theorem***
>
> Two compact surfaces are homeomorphic if and only if
>
>> **(a)** they have the same Euler characteristic, and
>> **(b)** they are both orientable or both nonorientable.

At the end of Chapter 3 we visualized orientable compact surfaces as spheres with some nice handles attached (zero handles in the case of the sphere). That number of handles (which can also be thought of as the number of "donut holes") for mT is m, and this number is called the *genus* of mT. Similarly, a nonorientable compact surface, mP, can be thought of as a sphere with m cross-caps attached. In this case, we will let this number m be what we mean by the *genus* of a nonorientable surface. "Genus" is often taken as a basic invariant of a compact surface, and the Euler characteristic is sometmes defined in terms of the genus. The relationship between genus and Euler characteristic is easy to see: for an orientable compact surface of genus g, the Euler characteristic is $\chi = 2 - 2g$; and for a nonorientable compact surface of genus g, the Euler characteristic is $\chi = 2 - g$.

• **Exercise 4.5** Determine all compact surfaces for which the genus equals the Euler characteristic.

The reader may find one final "visual" interpretation of the Euler characteristic interesting. As we have just recalled, any compact surface can be constructed by starting with a sphere, adding some handles (to form donut holes), and then adding some cross-caps. Adding one handle requires the removal of two open disks, before a cylindrical tube is attached, while adding a cross-cap requires removing just one open disk, before a Möbius band is attached. Initially the surface—a sphere—has Euler characteristic 2. Each handle added decreases the Euler characteristic by 2, and each cross-cap added further reduces the Euler characteristic by 1. Thus, the Euler characteristic of the final compact surface obtained in this way equals 2 minus the number of open disks removed during the construction of the surface.

Note that this "constructive" visual approach to the Euler characteristic involves intermediate punctured surfaces—after disks are removed and before attachments are made—just as occurred in the connected sum construction for compact surfaces addressed in Chapter 3. Again, it is important to remember that these intermediate "surfaces" are not compact surfaces since they fail to be 2-dimensional manifolds. Instead they are 2-dimensional manifolds with boundary, which were defined in Exercise S-2.4 in the Supplementary Exercises at the end of Chapter 2. In the previous paragraph, as the Euler characteristic was decreased appropriately with each attachment, we did not associate a number as an Euler characteristic for these intermediate surfaces. However, an Euler characteristic can be defined for compact, connected 2-dimensional manifolds with boundary, and an investigation of this notion is requested in Exercise S-4.11 at the end of this chapter.

4.2   PATTERNS, COMPLEXES, AND REGULARITY

There are many interesting and important consequences of the formula $V - E + F = \chi(X)$. In this section we will investigate some such consequences that involve the different kinds of faces and vertices a tiling on a compact surface may possess. As we develop these ideas, we will restrict our attention to tilings satisfying some (not too restrictive) properties that help us avoid some technical difficulties. Before doing so, however, we will develop a few computational ideas which apply to tilings in general.

If we look just within some small neighborhood of a vertex of a tiling on a compact surface, we will see segments of the edges ending there. We will designate the number of such edge segments we see as the *valence* of the vertex. Some caution should be observed when we consider the valence of a vertex in a tiling, since there could be some edge that starts and ends at the vertex. In other words, the valence of a vertex could "count" some edges twice, and the number of distinct edges ending at the vertex could be less than the valence of the vertex. For example, the tiling of the torus induced by the standard word $aba^{-1}b^{-1}$ possesses just one vertex, and it has valence 4 even though only two distinct edges end at that vertex. Similarly, the tiling of the projective plane induced by the standard word aa has just one vertex that has valence 2 even though the tiling has only one edge. However, note that each of the two vertices of the tiling of the sphere induced by its standard word aa^{-1} has valence 1; they are the two distinct endpoints of the single edge of the tiling.

Definition

For a tiling on a compact surface, and for integers $a \geq 1$ and $b \geq 1$

F_a denotes the number of faces of the tiling that are a-gons, and
V_b denotes the number of vertices of the tiling with valence b.

Note that for any tiling on a compact surface the numbers F_a $(a \geq 1)$ must sum to F, the total number of faces, and the numbers V_b $(b \geq 1)$ must sum to V, the total number of vertices.

EXAMPLE 4.11

The tiling labeled (b) on the sphere shown in Figure 4.1 possesses nine faces—five squares and four triangles. So for this tiling, $F_4 = 5$, $F_3 = 4$, and $F_a = 0$ for any a other than 3 or 4. This tiling can easily be checked to see that there are four vertices of valence 3 and five vertices of valence 4 so that $V_3 = 4$, $V_4 = 5$, and $V_b = 0$ for any b other than 3 or 4.

• **Exercise 4.6**

(a) For the tiling on the Klein bottle that was considered in Example 4.2, determine the numbers F_a $(a \geq 1)$ and V_b $(b \geq 1)$.

(b) For the tiling on the projective plane that was considered in Exercise 4.1, determine the numbers F_a ($a \geq 1$) and V_b ($b \geq 1$).

There is a beautiful and simple relationship among all the numbers V_b ($b \geq 1$) for a tiling. Imagine that V observers are stationed one at each of the V vertices of the tiling. Since each vertex of valence b is attached to b edge segments, an observer standing on the surface at such a vertex can "report" seeing b edges. Now, for a fixed $b \geq 1$, all together the V_b observers, stationed one at each of the V_b vertices of valence b, will report seeing bV_b edges. But since each edge has two end segments, each of the E edges of the tiling will be reported exactly twice among the whole group of the V observers. (If both end segments of some edge are attached to the same vertex, then the observer there counts that edge twice.) Thus, summing up all the numbers bV_b ($b \geq 1$) must yield $2E$.

Similarly, suppose that there are F observers stationed one on each of the F faces of the tiling. Then, for a fixed $a \geq 1$, each of the F_a observers stationed on an a-gon face will report seeing a edges surrounding the face, and together all F_a observers report seeing aF_a edges. But since there is a face on each side of each of the E edges of the tiling (perhaps the same face, if it meets itself along that edge), each edge is reported twice among the whole group. Thus, summing all the numbers aF_a ($a \geq 1$) also yields $2E$.

The results of the last two paragraphs are interesting and useful enough to summarize as a theorem.

Theorem

For a tiling on a compact surface,

$$2E = 1V_1 + 2V_2 + 3V_3 + 4V_4 + 5V_5 + \dots \text{ and}$$
$$2E = 1F_1 + 2F_2 + 3F_3 + 4F_4 + 5F_5 + \dots .$$

• **Exercise 4.7** Confirm that the equations of the theorem hold for each of the following tilings:

(a) the tiling on K considered previously in Example 4.2 and Exercise 4.6.
(b) the tiling on P considered previously in Exercises 4.1 and 4.6.
(c) the tiling on K induced by the word $abb^{-1}cadc^{-1}d$.

The first special type of tiling we will consider is called a *pattern*.

Definition

A *pattern* on a compact surface is a tiling on the surface such that

(a) no edge meets itself at its endpoint vertices, and
(b) no face meets itself along an entire edge.

The main motivation for introducing this special type of tiling is to introduce a condition that forces valence to equal the number of distinct edges ending at a vertex. Indeed condition (a) for a pattern ensures that all of the edges that meet at a vertex are distinct, and it also disallows 1-gon faces. Condition (b) is included as a condition for faces somewhat parallel to condition (a) for edges, although it still allows a face to meet itself, but only at vertices—not along any complete edge. Also condition (b) guarantees that at least two edges meet at each vertex, so that the valence of a vertex of a pattern must be at least 2. Note that the tilings considered in Example 4.11 and Exercise 4.6 are patterns, while no tiling induced by a plane model of a compact surface can be a pattern.

The equations given in the previous theorem yield relationships between the Euler characteristic and the numbers V_b and F_a for a pattern, as a little computation shows. For a pattern on a compact surface X,

$$
\begin{aligned}
2\chi(X) &= 2V - 2E + 2F \\
&= 2(V_2 + V_3 + V_4 + V_5 + \ldots) - (2F_2 + 3F_3 + 4F_4 + 5F_5 + \ldots) \\
&\quad + 2(F_2 + F_3 + F_4 + F_5 + \ldots) \\
&= (2V_2 + 2V_3 + 2V_4 + 2V_5 + \ldots) - (F_3 + 2F_4 + 3F_5 + \ldots)
\end{aligned}
$$

and also

$$
\begin{aligned}
2\chi(X) &= 2V - 2E + 2F \\
&= 2(V_2 + V_3 + V_4 + V_5 + \ldots) - (2V_2 + 3V_3 + 4V_4 + 5V_5 + \ldots) \\
&\quad + 2(F_2 + F_3 + F_4 + F_5 + \ldots) \\
&= (2F_2 + 2F_3 + 2F_4 + 2F_5 + \ldots) - (V_3 + 2V_4 + 3V_5 + \ldots).
\end{aligned}
$$

Moreover, adding these two equations produces the beautiful equation

$$
\begin{aligned}
4\chi(X) &= (2V_2 + V_3 - V_5 - 2V_6 - 3V_7 - 4V_8 - \ldots) \\
&\quad + (2F_2 + F_3 - F_5 - 2F_6 - 3F_7 - 4F_8 - \ldots)
\end{aligned}
$$

that is completely symmetric in the V_b's and the F_a's.

These results enable us to determine information about the types of patterns that can actually occur on various compact surfaces.

EXAMPLE 4.12

Consider any pattern on the sphere or the projective plane. Since the Euler characteristic of these surfaces is positive, so is $(2V_2 + V_3 - V_5 - 2V_6 - 3V_7 - 4V_8 - \ldots) + (2F_2 + F_3 - F_5 - 2F_6 - 3F_7 - 4F_8 - \ldots)$. Thus, one of the numbers V_2, V_3, F_2, or F_3 must be nonzero. Therefore, any pattern on S or P contains either a "small" face (a 2-gon or a 3-gon) or a "small" vertex (one of valence 2 or 3).

EXAMPLE 4.13

Now consider any pattern on the torus or the Klein bottle, both of which have Euler characteristic 0. In this case the value of $(2V_2 + V_3 - V_5 - 2V_6 - 3V_7 - 4V_8 - \dots) + (2F_2 + F_3 - F_5 - 2F_6 - 3F_7 - 4F_8 - \dots)$ must be 0 too. If it happens that $V_2 = V_3 = F_2 = F_3 = 0$, then all the terms in this expression must equal 0, and this says that $V = V_4$ and $F = F_4$. Therefore, if a pattern on T or K does not have a "small" face or a "small" vertex, then it consists entirely of 4-gon faces and vertices of valence 4.

• **Exercise 4.8**   A pattern on a compact surface X is made up of equal numbers of 3-gons and 4-gons and has eight vertices: five of valence 3, two of valence 4, and one of valence 5.

(a) Determine, with proof, the common number of 3-gons and 4-gons, and also determine the compact surface X.

(b) Sketch a picture showing such a pattern on a model of X in 3-space.

The conclusions of Examples 4.12 and 4.13, as well as the identities leading to them, appear in [**Firby**] where they are discussed for a more restrictive sort of pattern called a *complex*. This is the next special type of tiling we will discuss. The geometrical objects that complexes on compact surfaces emulate are called *polyhedra*. The faces of a *polyhedron* consist of actual rigid, flat polygons with edges that are straight line segments joining the vertices. Any face of such a polyhedron must have at least three edges, so polyhedra represent patterns without any 2-gon faces. And at any vertex of a polyhedron at least three edges meet, so vertices have valence at least 3. Moreover, two different polygonal faces of a polyhedron cannot meet each other twice; for if they did—either along the entire lengths of two different edges or at two vertices that do not lie on a common edge—then the faces must meet in their interiors. We will define a *complex* on a compact surface to be a pattern that satisfies these properties of polyhedra.

Definition

A *complex* on a compact surface is a pattern on the surface

(a) that contains no 2-gon faces and no vertices of valence 2,

(b) in which no face meets itself, not even at a vertex, and

(c) in which two distinct faces can meet at most once—either at a single vertex or along an entire single edge.

Note that all the results derived previously for patterns of course hold for complexes.

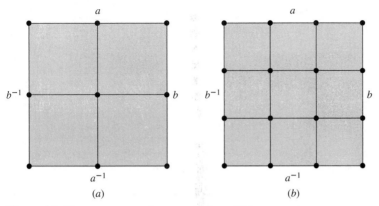

Figure 4.7 Two patterns on the torus. Pattern (b) is a complex.

EXAMPLE 4.14

Consider the two patterns shown in Figure 4.7 on the usual plane model for T. Note that both contain only 4-gon faces and only vertices of valence 4, as must be the case if no faces with three or fewer edges or vertices of valence 3 or less occur. However, the pattern labeled (a) is not a complex since each of its four faces meets each of the others twice. (The reader may want to check each pair of faces to see the ways in which they meet.) On the other hand, pattern (b), which has nine faces, is a complex.

EXAMPLE 4.15

Figure 4.8 shows a complex on the plane model for P consisting of six 5-gon faces (i.e., pentagons), ten vertices—each of valence 3, and fifteen edges. The edges of this complex are drawn as solid lines and curves and include edges that cover the edges of the plane model. The dashed lines shown on the same figure indicate edges of another complex on the plane model of P. This new complex is obtained from the first by placing a new vertex (indicated by an open dot) in the interior of each face of the original complex. If two new vertices lie in interiors of faces that share an edge in the original complex, then they are connected by a new edge crossing that edge of the original complex. This construction yields a complex on P consisting of ten 3-gon faces (i.e., triangles), six vertices—each of valence 5, and fifteen edges.

The new complex constructed in the last example is called the *dual* of the original complex. This dual construction can also be carried out for a pattern on a compact surface, but we will consider it in the context of complexes.

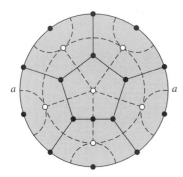

Figure 4.8 A complex on P and its dual complex.

Definition

Given a complex on a compact surface X, its *dual complex* is the complex on X constructed as follows:

(a) Place a new vertex in the interior of each face.
(b) Connect two new vertices lying in faces that share an edge with a new edge drawn crossing the edge shared by those faces.

Note that the number of edges is the same for the original complex and its dual complex, while the number of vertices and the number of faces are interchanged. Moreover, the number of a-gon faces of the original complex equals the number of vertices of the dual complex with valence a, and the number of vertices of the original complex with valence b equals the number of b-gon faces of the dual complex. Clearly the dual complex of the dual complex of an original complex corresponds to the original complex.

• **Exercise 4.9** A complex on the sphere is shown in Figure 4.9, first as the surface of a tetrahedron with one corner truncated and second on the usual 2-gon plane model for S.

(a) For this complex, determine all the numbers F_a ($a \geq 3$), V_b ($b \geq 3$), V, E, and F.
(b) Copy the plane model for the given complex, and then draw its dual complex on the plane model.

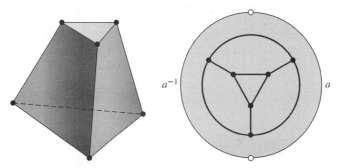

Figure 4.9 The complex on S for Exercise 4.9.

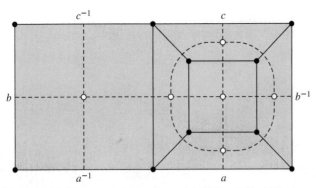

Figure 4.10 The regular complexes $\{4, 3\}S$ and $\{3, 4\}S$ are duals of each other.

(c) For the dual complex, determine the numbers F_a ($a \geq 3$), V_b ($b \geq 3$), V, E, and F, and state the relationship between these numbers and those for the original complex.

The complexes considered in Examples 4.14 and 4.15 are of a very special type. They are *regular* complexes.

Definition

A complex on a compact surface X is called *regular* if there are integers $a \geq 3$ and $b \geq 3$ such that each face of the complex is an a-gon and each vertex of the complex has valence b.

Such a regular complex is said to be a *regular complex of type* $\{a, b\}$ on X or, more briefly, a *type* $\{a, b\}X$ *regular complex.*

The complex of Example 4.14 on T is a regular complex of type $\{4,4\}T$. In Example 4.15, the complex on P introduced first is a regular complex of type $\{5,3\}P$ and its dual complex, also introduced in the example, is a regular complex of type $\{3,5\}P$.

EXAMPLE 4.16

The complex on the sphere given by the cube surface is a regular complex of type $\{4, 3\}S$. The dual complex on S will be a regular complex of type $\{3, 4\}S$ with eight triangular faces, six vertices with valence 4, and twelve edges. Figure 4.10 shows these two regular complexes drawn on the 6-gon plane model for S that appeared previously in Figure 4.6. Recall that this plane model was *constructed* from the cube by slicing along three edges.

The regular complex types $\{4, 3\}S$ and $\{3, 4\}S$ represent two of the five famous regular spherical polyhedra—or *Platonic solids*—that were known in an-

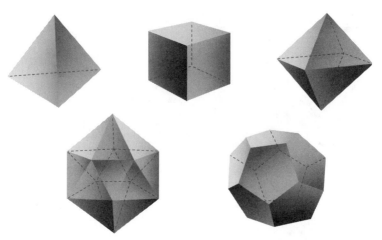

Figure 4.11 The five Platonic solids represent regular complexes on S.

cient times. In addition to $\{4,3\}S$ (the cube) and $\{3,4\}S$ (the *regular octahedron*), these include the *regular tetrahedron* (represented by $\{3,3\}S$), the *regular icosahedron* (represented by $\{3,5\}S$), and the *regular dodecahedron* (represented by $\{5,3\}S$). These polyhedra are illustrated in Figure 4.11. Note that the prefixes *octa-, tetra-, icosa-,* and *dodeca-* refer to the numbers of faces that appear on these polyhedra, namely, eight, four, twenty, and twelve, respectively. Sometimes the prefix *hexa-* is used to refer to the numeral six, and the (six-faced) cube is called the *regular hexahedron*. Also note that $\{3,5\}S$ and $\{5,3\}S$ are dual complexes of each other, while $\{3,3\}S$ is "self-dual."

There is a straightforward approach to determining the possible regular complex types $\{a, b\}X$ which can occur on a compact surface X. Simple special cases of equations developed earlier lead us to conclude that $2E = aF$ and $2E = bV$ for such a regular complex. It follows that

$$ab\chi(X) = ab(V - E + F)$$
$$= a(bV) - abE + b(aF)$$
$$= 2aE - abE + 2bE$$
$$= 2E\left(a - \frac{ab}{2} + b\right)$$

Dividing by ab and rearranging yields $\chi(X) = 2E(\frac{1}{a} + \frac{1}{b} - \frac{1}{2})$. It now remains to determine the possible ordered pairs (a,b) of integers that can satisfy this equation. This task turns out to be especially easy when $\chi(X)$ is nonnegative! Noting that $2E$ is positive, we conclude that $\frac{1}{a} + \frac{1}{b} > \frac{1}{2}$ if $\chi(X)$ is positive (i.e., if $X = S$ or P) and that $\frac{1}{a} + \frac{1}{b} = \frac{1}{2}$ if $\chi(X) = 0$ (i.e., if $X = T$ or K).

Let us investigate each of the cases $X = S, P, T,$ and K separately.

For the sphere S, $\chi(S) = 2$ is positive, and the integer pairs (a, b) with $a \geq 3$ and $b \geq 3$ that solve the inequality $\frac{1}{a} + \frac{1}{b} > \frac{1}{2}$ are precisely $(a, b) = (3, 3)$, $(3, 4)$, $(3, 5)$, $(4, 3)$, and $(5, 3)$, which correspond to the five Platonic solids. Thus, all of these five possible regular complex types on S actually occur on S. Moreover, there is just one way that each can occur in the sense that the number

E of edges is completely determined by a and b in the relation $2 = \chi(S) = 2E(\frac{1}{a} + \frac{1}{b} - \frac{1}{2})$.

• **Exercise 4.10** Use the above relation along with the equations $aF = 2E$ and $bV = 2E$ to show that a regular complex on S of type $\{3,5\}S$ must have 20 faces, 12 vertices, and 30 edges. Based on this result, how many faces, edges, and vertices must a type $\{5,3\}S$ complex have?

Computations similar to that requested in Exercise 4.10 for the other possible types of regular complexes on S reveal that they too can occur only with certain numbers of edges, faces, and vertices. These results for the sphere are summarized in the following theorem.

Theorem

The regular complex types that can occur on S are

$\{3,3\}S$ with 4 faces, 4 vertices, and 6 edges;
$\{3,4\}S$ with 8 faces, 6 vertices, and 12 edges;
$\{4,3\}S$ with 6 faces, 8 vertices, and 12 edges;
$\{3,5\}S$ with 20 faces, 12 vertices, and 30 edges; and
$\{5,3\}S$ with 12 faces, 20 vertices, and 30 edges.

It is natural to next consider the case of regular complexes on P since it, similarly to S, has positive Euler characteristic. If $\{a,b\}P$ is a regular complex on P, then just as in the case of S, the possibilities for (a,b) are $(a,b) = (3,3)$, $(3,4)$, $(3,5)$, $(4,3)$, and $(5,3)$.

Now, we have already seen examples of type $\{3,5\}P$ and type $\{5,3\}P$ complexes on P in Example 4.15. Moreover, the relation $1 = \chi(P) = 2E(\frac{1}{a} + \frac{1}{b} - \frac{1}{2})$ with $(a,b) = (3,5)$ tells us that any type $\{3,5\}P$ complex must occur with $E = 15$ edges and, hence, with $F = 10$ triangular faces and $V = 6$ vertices. Then its dual complex, $\{5,3\}P$, occurs only with $E = 15$ edges, $V = 10$ vertices, and $F = 6$ pentagonal faces. Exercise S-4.7 in the Supplementary Exercises at the end of this chapter asks the reader to show that none of the possible regular complex types $\{3,3\}P$, $\{3,4\}P$, or $\{4,3\}P$ can occur as complexes on the projective plane.

Theorem

The regular complex types that can occur on P are

$\{3,5\}P$ with 10 faces, 6 vertices, and 15 edges; and
$\{5,3\}P$ with 6 faces, 10 vertices, and 15 edges.

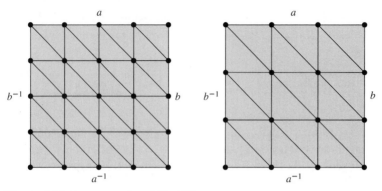

Figure 4.12 Two examples of $\{3,6\}T$.

The possible regular complex types $\{a,b\}X$, where X is either T or K, correspond to pairs (a,b) satisfying $\frac{1}{a} + \frac{1}{b} = \frac{1}{2}$. Now the solutions for (a,b) are $(a,b) = (4,4)$, $(3,6)$, and $(6,3)$. A complex of type $\{4,4\}T$ with 9 faces has already been seen in Example 4.14, and it is not difficult to construct other examples of type $\{4,4\}T$ complexes: on the usual rectangular plane model for T, use more than three rows or columns of square faces. Figure 4.12 shows several different examples of type $\{3,6\}T$ complexes. Examples of type $\{6,3\}T$ complexes may be obtained by forming the dual complex of any of the $\{3,6\}T$ examples, but a famous example—introduced in the late 1800s by Heawood—is shown in Figure 4.13. The Heawood model of $\{6,3\}T$ uses exactly 7 faces (the minimum number that could be used for such a complex) and has the property that each face meets all of the other six faces. Note that the plane model shown in the figure is the usual plane model for T displayed as an interesting parallelogram cut out of a tessellation of the plane by regular hexagons.

The situation on K is similar with just a few differences. If one looks at the examples of $\{4,4\}T$ and $\{3,6\}T$ shown in Figures 4.7 and 4.12 and changes the gluing direction of the left edge of each plane model, then what occurs are examples of complexes of types $\{4,4\}K$ and $\{3,6\}K$ on the usual plane model of K. Then, of course, we know that $\{6,3\}K$ can occur on K since it is the dual of

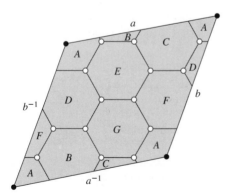

Figure 4.13 Heawood's model of $\{6,3\}T$ using just seven hexagons.

$\{3, 6\}K$. However, the conditions of the Heawood model of $\{6, 3\}T$ cannot be translated to the case of the Klein bottle; that is, $\{6, 3\}K$ cannot occur as a complex on K with only 7 faces. This will follow from a wonderful result that will appear in the next section.

Note that the basic relation $0 = \chi(X) = 2E(\frac{1}{a} + \frac{1}{b} - \frac{1}{2})$, for $X = T$ or K, allows infinitely many solutions for E given a and b. Thus, it is not surprising that there are many—in fact infinitely many—different ways for each of the possible regular complex types on T and K to occur.

Theorem

If $X = T$ or K, then the regular complex types that can occur on X are $\{4, 4\}X$, $\{3, 6\}X$, and $\{6, 3\}X$.

These regular complex types on T and K can occur in infinitely many different ways.

The more modern history of regular polyhedra-like structures on general compact surfaces extends at least back into the late nineteenth century when Felix Klein described a type $\{7, 3\}3T$ regular complex. The problem of existence of regular complexes on a compact surface X corresponding to all possible positive integral solutions of $V - E + F = \chi(X)$, $aF = bV = 2E$ was considered more recently. Special cases of this problem were solved by Errera in the 1920s and Threlfall in the 1930s, and the general result was finally verified for regular *patterns* in the 1980s by Edmonds, Ewing, and Kulkarni [**Edmonds**]—with the exception of $\{3, 3\}P$, which cannot occur even as a regular pattern on the projective plane.

• **Exercise 4.11**   It is a fact that the regular complex type $\{4, 5\}7T$ does occur as a regular complex on $7T$. Determine V, E, and F for this regular complex.

A few final comments about the notion of regularity of a complex are necessary. Our definition of regular complex reflects a "local" idea of uniform behavior of the complex: if we stand inside any face we see the same number of edges surrounding us and if we stand on any vertex we see the same number of edges ending there. Mathematicians usually look at regularity from a "global" point of view as well. The Platonic solids, which we now know correspond to the regular complexes on S, possess a great deal of symmetry. Especially of interest are the possible permutations of the vertices, edges, and faces that preserve "incidence" (how they are connected to each other)—such permutations are called the *automorphisms* of the complex. The symmetry of the Platonic solids is determined by the existence of certain types of automorphisms, and often the definition of regularity of complexes on other compact surfaces includes a requirement that these certain types of automorphisms exist for the complex under consideration. This extra requirement is not satisfied by some of the regular complexes we have already seen, especially those on the Klein bottle. These ideas are closely related

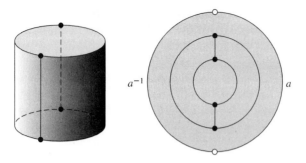

Figure 4.14 A map shown on S and also on a plane model for S.

to the part of abstract algebra called group theory, and regular complexes on compact surfaces are of much interest to mathematicians who work in this area.

4.3 COLORING MAPS ON SURFACES

Imagine a map (in the geographic sense) sketched on the surface of the sphere. The boundary curves of the map divide the surface into regions that we may think of as countries. An example of such a map, drawn on the sphere itself and also on a plane model of S, is shown in Figure 4.14. The boundary curves of this map divide the surface into four countries, two 2-gons and two 4-gons; and three edges meet at each of the four vertices. Now if we wish to color the four countries in such a way that any two countries sharing an edge are colored differently, then certainly four colors would do the job since each country could be colored with a different color. However, this task could be accomplished with fewer colors, namely, three colors for the map we are considering: color the two 2-gons blue, say, and use red and green to color the two 4-gons. However, two colors would not suffice for our example since there is a set of three countries, each pair of which share an edge.

The previous example illustrates the notion we will consider in this section, determining the number of colors required to color maps on various surfaces. The associated ideas have a rich history in mathematics and include the famous "Four Color Problem," which dates back to the nineteenth century and asks if four colors suffice for coloring all maps on the sphere. It is remarkable that this question was finally answered affirmatively in the 1970s. Our treatment of coloring results follows the usual approach offered in other modern day textbooks such as [**Firby**] or [**Henle**]. For a very thorough survey—and a complete bibliography on coloring results—the reader is encouraged to read S. Stahl's wonderful 1985 article "The Other Map Coloring Theorem" [**Stahl**].

To make things a little more precise, we must first say officially what a *map* is. Our example motivates us to define a map to be some kind of pattern (but not necessarily a complex) in which the faces play the role of the countries. Now even though some geographic maps may exhibit boundaries of countries that contain corner "vertices" of valence 2 (such as the southwest corner of Wyoming on a map of the United States), such vertices play no role in our coloring problem and can be deleted from consideration. This extra condition is all that we require in the definition of a map on a compact surface.

Definition

A *map* on a compact surface X is a pattern on X in which each vertex has valence at least 3.

For a positive integer N, a map on a compact surface is said to be *N-colorable* if, given N different colors, each face of the map can be colored with one of the colors in such a way that any two distinct faces sharing an edge are colored with different colors.

According to this definition, the example we have seen is a map on S that is 3-colorable but not 2-colorable. It is, by the way, also 4-colorable and also 5-colorable and so on; the definition does not require that *all* N colors need to be used in order for a map to be N-colorable. Here are some general (and useful) facts that are quite obvious:

Remarks

(a) If N_1 and N_2 are positive integers with $N_1 < N_2$ and a map on a compact surface is N_1-colorable, then it is also N_2-colorable.

(b) If a map on a compact surface has F faces, then it is F-colorable.

EXAMPLE 4.17

Three regular complexes we discussed in the previous section provide interesting examples. A regular complex on S of type $\{3, 3\}S$ contains exactly four triangular faces, any two of which share an edge. This map is, therefore, 4-colorable but is not N-colorable for any N less than 4. Similarly, the Heawood model of a regular complex on T of type $\{6, 3\}T$ (refer to Figure 4.13) provides an example of a map on T that contains exactly seven faces, any two of which share an edge, and this map on T is 7-colorable but is not N-colorable for any N less than 7. And also, the regular complex on P of type $\{5, 3\}P$ contains six pentagonal faces, any two of which share an edge, so that this map proves to be 6-colorable but not N-colorable for any N less than 6.

• **Exercise 4.12**

(a) A nine-faced map on the plane model of the torus is shown on the left in Figure 4.15. Find a coloring of this map that uses as few colors as possible.

(b) A nine-faced map on the plane model of the Klein bottle is shown on the right in Figure 4.15. Find a coloring of this map that uses as few colors as possible.

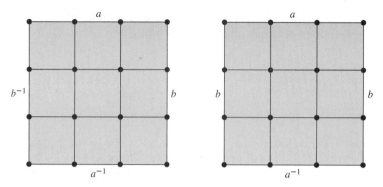

Figure 4.15 The maps on T and K for Exercise 4.12.

To obtain some useful general results, we will begin with a computation involving some familiar formulas. Consider any map on a compact surface X. For such a map,

$$
\begin{aligned}
2E &= 3V_3 + 4V_4 + 5V_5 + \dots \\
&\geq 3V_3 + 3V_4 + 3V_5 + \dots \\
&= 3(V_3 + V_4 + V_5 + \dots) \\
&= 3V
\end{aligned}
$$

so that $V \leq 2E/3$. It follows that

$$
\begin{aligned}
\chi(X) &= V - E + F \\
&\leq \frac{2E}{3} - E + F \\
&= F - \frac{E}{3}
\end{aligned}
$$

whence $E \leq 3(F - \chi(X))$. Upon multiplying by $2/F$, we obtain

$$
\frac{2E}{F} \leq 6\left(1 - \frac{\chi(X)}{F}\right).
$$

The number $2E/F$ can be thought of as the "average number of edges per face" for the map since $2E$ equals the number of edges of the map "counted face by face."

Now suppose that N is a positive integer. We are now in a position to prove that if $6(1 - \chi(X)/F) < N$ for all maps on X with more than N faces, then all maps on X are N-colorable. We will use the Principle of Mathematical Induction to do this, inducting on the number F of faces of maps on X.

Certainly all maps on X that satisfy $F \leq N$ are N-colorable. So the "initial case," which may be thought of as including all $F \leq N$, is certainly clear. So let k be a fixed but arbitrary integer with $k \geq N$, and assume that all maps on X with $F = k$ faces are N-colorable. Consider a map on X with $F = k + 1$ faces. Since

$k + 1 > N$, we know that $6(1 - \chi((X)/F)) < N$, and we have already seen that also $2E/F \leq 6(1 - \chi((X)/F))$ holds. So it follows that $2E/F < N$. This forces at least one face of the map to have fewer than N edges. Indeed, otherwise, we would have

$$2E = NF_N + (N + 1)F_{N+1} + (N + 2)F_{N+2} + \ldots$$
$$\geq NF_N + NF_{N+1} + NF_{N+2} + \ldots$$
$$= N(F_N + F_{N+1} + F_{N+2} + \ldots)$$
$$= NF$$

which says that $2E/F \geq N$, contrary to $2E/F < N$. So let D denote a face of our considered map with a minimal number, say a, of edges. Then $a < N$. Now we can form a new map on X with just k faces by collapsing D so that all of its edges and vertices shrink to a point on X. (Note that the tiling obtained will be a pattern in which some faces may meet themselves but only possibly at vertices, not along any complete edges.) The new map created is N-colorable by the induction hypothesis, and any appropriate coloring of it using N colors induces a coloring of all the faces of the original map, except for D. However, since the a faces of the original map that share an edge with D use up at most a colors and $a < N$, there is at least one of the N colors available to color D. Hence, the original map with $F = k + 1$ faces is indeed N-colorable.

So the Principle of Mathematical Induction yields the following result.

Theorem

Let N be a positive integer, and let X be a compact surface. If all maps on X with more than N faces satisfy $6(1 - \frac{\chi(X)}{F}) < N$, then all maps on X are N-colorable.

It is very easy to apply this result to compact surfaces with nonnegative Euler characteristic, namely S, P, T, and K. Since both S and P have positive Euler characteristic, $N = 6$ satisfies $6(1 - \frac{\chi(X)}{F}) < N$ for all maps on those surfaces. And since $\chi(T) = \chi(K) = 0$, $N = 7$ works for maps on T and K.

Corollary

Any map on S or P can be colored with 6 or fewer colors.
Any map on T or K can be colored with 7 or fewer colors.

If we define the *chromatic number* of a map to be the smallest integer N such that the map is N-colorable, then these results specify upper bounds for the chromatic numbers of maps on each of the surfaces considered. It is, of course, of interest to determine whether these upper bounds are in fact least upper bounds—in this case maxima—of chromatic numbers for maps on the surfaces. This maximum of the chromatic numbers of maps on a compact surface X will equal the minimum number of colors required to color all maps on X, and we will define this number to be the *coloring number* of the surface X.

Definition

For a compact surface X, the *coloring number* of X is the maximum of the chromatic numbers of maps on X (i.e., the minimum number of colors required to color all maps on X).

We have already seen that the regular complex of type $\{5, 3\}P$ with six faces requires six colors and that the regular complex of type $\{6, 3\}T$ with seven faces requires seven colors. Thus, the coloring numbers of P and T are 6 and 7, respectively. However, the cases of the sphere and the Klein bottle are much different. For the sphere, it was proven, by Heawood in 1890, that any map can be colored by five or fewer colors and, finally in 1977 by Appel and Haken, that just four colors suffice to color all maps on S. Given an example, say $\{3, 3\}S$, of a map on S requiring four colors, we have the result that the coloring number of the sphere is 4. For the Klein bottle, the less famous, but certainly not less important, coloring result was obtained in 1934 when Franklin proved that six colors suffice to color all maps on K. A modification of the Heawood model of the complex of type $\{6, 3\}T$ on the torus yields a map on K requiring all six colors. It is shown in Figure 4.16 where the seven faces are labeled A through G. If one assigns colors 1 through 6 beginning with face A and continuing in alphabetical order through face F, it will be clear that all six colors must be used for these six faces. But then the seventh face G that remains may be assigned either color 5 or color 6. So the coloring number of the Klein bottle is 6. Summarizing, we have the following theorem.

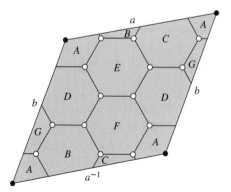

Figure 4.16 A map on K requiring 6 colors.

> **Theorem**
>
> The coloring numbers of the four compact surfaces with nonnegative Euler characteristic are
>
> 4 for the sphere S,
> 6 for the projective plane P,
> 7 for the torus T, and
> 6 for the Klein bottle K.

When Heawood proved the 5-coloring result for maps on the sphere in 1890, he also was able to obtain an upper bound for chromatic numbers of maps on compact surfaces of negative Euler characteristic. This is a result that we can derive easily from our previous calculations. Suppose that X is a compact surface with $\chi(X) < 0$. Let N be a positive integer and suppose that a map on X has more than N faces, so that $F \geq N + 1$. Then, $\frac{1}{F} \leq \frac{1}{N+1}$ and, since $\chi(X) < 0$, $\frac{\chi(X)}{F} \geq \frac{\chi(X)}{N+1}$. It follows that $1 - \frac{\chi(X)}{F} \leq 1 - \frac{\chi(X)}{N+1}$ and then that $6(1 - \frac{\chi(X)}{F}) \leq 6(1 - \frac{\chi(X)}{N+1})$. Now if N exceeds the right hand side $6(1 - \frac{\chi(X)}{N+1})$, then N will exceed the left hand side $6(1 - \frac{\chi(X)}{F})$ for all maps on X with more than N faces and this will assure us that all maps on X are N-colorable.

To explicitly determine the values of such integers N, we can rearrange the inequality $6(1 - \frac{\chi(X)}{N+1}) < N$ to obtain $6(N + 1 - \chi(X)) < N(N + 1)$, which yields the quadratic inequality $N^2 - 5N + 6\chi(X) - 6 > 0$. A simple application of the quadratic formula shows that the positive integer values of N that satisfy this inequality must exceed $\frac{5 + \sqrt{49 - 24\chi(X)}}{2}$ and the least of those values of N is given explicitly by $\left\lfloor \frac{7 + \sqrt{49 - 24\chi(X)}}{2} \right\rfloor$, where $\lfloor \ldots \rfloor$ denotes the "floor" or "greatest integer" function. This number will be designated as the *Heawood number* and will be denoted by $H(X)$.

> **Definition**
>
> Let X be a compact surface with negative Euler characteristic. The *Heawood number* of X is defined to be
>
> $$H(X) = \left\lfloor \frac{7 + \sqrt{49 - 24\chi(X)}}{2} \right\rfloor.$$

We have proven the following theorem.

> **Theorem**
>
> Let X be a compact surface with negative Euler characteristic. Any map on X is $H(X)$-colorable. Thus, $H(X)$ is an upper bound for the chromatic numbers of maps on X.

• **Exercise 4.13** Show that all maps on $23T$ are 23-colorable. Will fewer than 23 colors suffice to color all maps on $23T$?

It is interesting to note that, although the results above were derived for compact surfaces with negative Euler characteristic, the formula for $H(X)$ can be evaluated for compact surfaces X with Euler characteristic of 2, 1, or 0 to obtain $H(S) = 4$, $H(P) = 6$, $H(T) = 7$, and $H(K) = 7$. Thus, in these cases too, the Heawood number provides an upper bound for the chromatic numbers of maps on each of the basic surfaces! The table below shows these and a few other values of the Heawood numbers.

X	$\chi(X)$	$H(X)$
S	2	4
P	1	6
T or $2P$ (K)	0	7
$3P$ $(P\#T)$	-1	7
$2T$ or $4P$ $(K\#T)$	-2	8
$5P$ $(P\#2T)$	-3	9
$3T$ or $6P$ $(K\#2T)$	-4	9
$7P$ $(P\#3T)$	-5	10

Notice that the Heawood number $H(X)$ in fact provides the coloring numbers when $X = S$, P, and T but fails to do so for $X = K$. In fact, in 1890 Heawood conjectured for all orientable compact surfaces, other than the sphere, that indeed $H(X)$ gives the coloring number. (At that time, the "four color theorem" for the sphere seemed so far out of reach that it was not easy to believe in!) Similarly in 1912 Tietze made the analogous conjecture concerning all nonorientable surfaces. (Remember, the case in which we know this conjecture would fail—for the Klein bottle—was not settled until 1934.) It is remarkable that all cases of this "Heawood-Tietze Conjecture," with the exception of the Klein bottle case, were verified before 1970 and, of course, the (unconjectured) case of the sphere was verified finally in 1977.

Theorem

For any compact surface X, other than K, the coloring number of X is given by the Heawood number $H(X)$. The coloring number of K is 6, while $H(K) = 7$.

• **Exercise 4.14** Determine the coloring number for each of the following compact surfaces.

(a) $P\#7T$

(b) $12P$

(c) $2P$

Recall the observation, made in the previous section, that the regular complex type $\{6,3\}K$ cannot be realized on the Klein bottle K using just seven hexagonal faces. This can now be confirmed using the 1934 result of Franklin. Indeed, if a regular complex of type $\{6,3\}K$ did occur with seven hexagonal faces, then each such face would share an edge with each of the other six faces. Such a complex would be an example of a map on K requiring seven colors, contrary to Franklin's theorem.

4.4 CHAPTER SUMMARY

A tiling on a compact surface generalizes the notion of the surface of a polyhedron, and in this chapter the Euler polyhedral formula $V - E + F = 2$ for the sphere led us to the notion of the Euler characteristic as an invariant of compact surfaces. To obtain a device for computing the Euler characteristic, we appealed to plane models, which themselves provide special–one-faced–tilings. Further, when combined with orientability, the Euler characteristic served to distinguish all the normal form surfaces mentioned in the classification theorem of Chapter 3, and we were able to show that two compact surfaces are homeomorphic if and only if they have the same Euler characteristic and orientability.

By restricting the types of tilings under consideration, we were able to prove many interesting and useful facts about patterns and complexes on compact surfaces, even from very little information. For example, we were able to determine all possible types of regular complexes on the four compact surfaces with nonnegative Euler characteristic and, in particular, prove the classical result that the surfaces of the five Platonic solids exhaust all such possibilities on the sphere. Also, maps on compact surfaces were introduced as certain kinds of patterns, and the notion of colorability of maps led to a complete investigation of the coloring numbers for all compact surfaces, culminating with discussion of the Heawood-Tietze conjecture and the famous four color theorem for maps on the sphere.

In the next chapter we will turn our attention to the theory of graphs: sets of vertices connected by edges. There compact surfaces will appear as surfaces in which graphs may lie without edges crossing unnecessarily. And the Euler characteristic will again play an important role.

4.5 SUPPLEMENTARY EXERCISES

Exercise S-4.1 Show that if X and Y are compact surfaces, then $\chi(X\#Y) = \chi(X) + \chi(Y) - 2$. Do this by examining, in a sequence of sketches, how plane models for X and Y combine to form a plane model for $X\#Y$.

Exercise S-4.2 Use the result of the previous exercise to determine all compact surfaces X for which:

(a) $\chi(P\#X) \cdot \chi(K\#X) = 0$
(b) $\chi(P\#X) \cdot \chi(K\#X) = 72$
(c) $\chi(P\#X) \cdot \chi(K\#X) = 30$

Exercise S-4.3 Use the Euler characteristic to show that no two different alternative nonorientable normal form surfaces (i.e., of the form $K \# mT$ $(m \geq 0)$ or $P \# mT$ $(m \geq 0)$) can be homeomorphic. (Note that there are actually *three* cases to consider here!)

Exercise S-4.4

(a) If there exists a tiling on a compact surface X with exactly 24 vertices, exactly 50 edges, and exactly 21 faces, then what must the surface be?

(b) If there exists a tiling on a compact surface X with exactly 12 vertices, exactly 14 edges, and exactly 5 faces, then what must the surface be?

Exercise S-4.5 Show that any pattern on a compact surface of negative Euler characteristic (i.e., one other than S, T, K, or P) must contain either a face with at least five edges or a vertex of valence at least 5.

Exercise S-4.6 We will say that a face of a tiling on a compact surface is an "odd face" if it is an a-gon for some odd integer $a \geq 1$. Prove that, for any tiling on a compact surface, the number of odd faces is even.

Exercise S-4.7

(a) Show that the regular complex types $\{3, 3\}P$, $\{3, 4\}P$, and $\{4, 3\}P$ cannot occur as complexes on the projective plane.

(b) On a plane model for P, draw an example of a *pattern* all of whose faces are 3-gons and all of whose vertices have valence 4.

Exercise S-4.8 Although we have discussed regularity in this chapter only for complexes, we can also consider a pattern on a compact surface to be *regular* if there are some integers $a \geq 2$ and $b \geq 2$ such that all faces are a-gons and all vertices have valence b.

(a) Show that regular patterns with 2-gon faces can occur only on the sphere or the projective plane.

(b) Determine all possibilities for regular patterns with 2-gon faces on the sphere, and draw some sketches to show that they all actually occur.

Exercise S-4.9 Find the least positive integer n for which the following statement is false: "$n + 6$ *colors are required to color some map on* nT."

Exercise S-4.10 Suppose that X is a compact surface, that 72 colors are required to color some map on X, and that all maps on X are 75-colorable. Determine the number of different, up to homeomorphism, possibilities for the compact surface X that could satisfy the given conditions.

Exercise S-4.11 The definition of a 2-dimensional manifold with boundary was presented in Exercise S-2.4 in the Supplementary Exercises at the end of Chapter 2. Let us now define a *bordered surface* to be a connected 2-dimensional manifold with boundary.

(a) Consider some examples of compact bordered surfaces that are not compact surfaces. On each, or on a plane model, draw some tilings and, based on your choices of tilings, conjecture what the Euler characteristic would be for each compact bordered surface you consider.

(b) What numbers do you think arise as Euler characteristics of compact bordered surfaces? Explain your answer in terms of either plane models or space models for these objects.

Chapter **5**

Graphs and Topology

$\mathbf{A}$ graph can be thought of as a finite set of points in space along with some arcs connecting some of those points. As such, a graph is just an object—that is, a set of points in space—whose topological properties we may study. In fact, graphs have arisen already in our study of tilings on compact surfaces: given a complex on a surface, its vertices and edges form a graph. Historically, graphs were among the first objects to which topological ideas were applied: the mathematical fields of topology and graph theory can be traced back to a common birth in the work of Leonard Euler in the 1700s. In their modern contexts, topology and graph theory share many ideas and problems, and each of these mathematical fields has been advanced by supporting work performed in the other.

In this chapter we will be introduced—in Section 1—to the basic ideas of the theory of graphs and then study—in Section 2—various kinds of "paths" in graphs. Readers who wish to obtain a more thorough understanding of these notions than our sections provide are referred to graph theory textbooks such as *Introduction to Graph Theory* by R. J. Wilson [**Wilson**]. In Section 3, the notions of "planarity" of graphs and embedding graphs in compact surfaces will be presented. These topics appear in a number of introductory books on graph theory (e.g., [**Wilson**]) or on topology (e.g., [**Firby**]); a thorough, advanced treatment may be found in *Graphs, Groups, and Surfaces* by A. White [**White**].

5.1 WHAT IS A GRAPH?

Although a visually motivating description of a graph was given above, it is useful now to give a more precise definition to begin our discussion of graph theory in its own right.

> **Definition**
>
> A *graph* consists of a finite set V, whose elements are called the *vertices* of the graph, and a set E of some two-element subsets of V. The elements of E are called the *edges* of the graph.

99

In this definition, the abstract idea of an edge—as a set of two verticies—incorporates our intuitive topological feeling that the edges connecting vertices of a graph should not depend on the actual arcs themselves as much as just the end-point vertices of the arcs. Thus, if we visualize an edge as an arc in space, that arc corresponds to the edge consisting of the two-element set of its endpoint vertices.

EXAMPLE 5.1

Let us start with a 3-element vertex set $V = \{a, b, c\}$. There are many edge sets that we may use to construct different graphs with vertex set V. The largest such edge set would contain all 2-element subsets of V, and so it would be $E = \{\{a, b\}, \{a, c\}, \{b, c\}\}$. Two other possible edge sets are $E = \{\{a, c\}, \{b, c\}\}$ or $E = \{\{a, b\}\}$. Even the empty set qualifies as a possible edge set! It should be easy for the reader to check that there are exactly eight different edge sets that could be associated with this set V of vertices.

A useful device for displaying a graph is called a *graph diagram*. A graph diagram for a graph employs dots in the plane to represent the graph's vertices and lines or arcs in the plane that connect the dots to represent the graph's edges. A line connecting the dots corresponding to two vertices is not allowed to pass through any other dot that represents any other vertex. Of course, a graph diagram just *represents* its graph—it is not the graph itself. Also, many graph diagrams can be found that represent a given graph.

Note that our definition of graph does not allow any "loops"—edges from a vertex to itself—nor does it allow more than one edge to connect a pair of vertices. These ideas are sometimes of interest in graph theory, as is the possibility of assigning a "direction" to each edge of a graph. But none of these conditions will be needed in the main part of our work with graphs in this chapter.

Figure 5.1 illustrates eight graph diagrams, one representing each of the eight graphs associated with the 3-element vertex set of the previous example. Among these eight diagrams, there are just four types of diagrams that "look different." In order to make this phenomenon more precise, it is convenient to introduce a notion of "equivalence" of graphs.

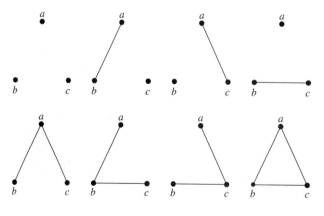

Figure 5.1 The eight graphs associated with the vertex set $V = \{a, b, c\}$.

> **Definition**
>
> Two graphs—G_1 with vertex set V_1 and edge set E_1 and G_2 with vertex set V_2 and edge set E_2—are called *isomorphic* if there is a bijection $f : V_1 \longrightarrow V_2$ such that $\{a, b\}$ belongs to E_1 if and only if $\{f(a), f(b)\}$ belongs to E_2.

In other words, we say that two graphs are isomorphic if there is a one-to-one correspondence between their vertex sets such that if any pair of vertices in either of the graphs is joined by an edge, then the corresponding pair of vertices in the other graph is also joined by an edge. So in Figure 5.1, there are just four graphs portrayed—up to isomorphism.

Among all graphs with an n-element vertex set V are two that are "extreme" in some sense. One is the graph with vertex set V and with empty edge set: it is called the *null graph* with n vertices. The other "extreme" graph is important enough to merit its own displayed definition.

> **Definition**
>
> The *complete graph* on n vertices is the graph with an n-element vertex set and whose edge set contains all of the $\binom{n}{2} = \frac{n(n-1)}{2}$ possible edges. The complete graph on n vertices is denoted by K_n.

EXAMPLE 5.2

Graph diagrams for the complete graphs K_3, K_4, and K_5 are shown in Figure 5.2. Note that some of the lines representing edges in the diagrams for K_4 and K_5 cross each other in the plane at points other than their endpoints.

• **Exercise 5.1**

(a) Try to draw some other graph diagrams for K_4 in which no arcs representing edges cross in the plane at points other than their endpoints.

Figure 5.2 Graph diagrams for some complete graphs.

(b) Try to draw some other graph diagrams for K_5 in which no arcs representing edges cross in the plane at points other than their endpoints. If you are not successful, try to find a diagram that leaves just one such "extra crossing."

Of course, a complete graph on n vertices has the maximum number of edges that *any* graph with n or fewer vertices could have. Moreover, such a complete graph actually contains isomorphic "copies" of all of those graphs as *subgraphs* in the sense of the following definition.

Definition

A graph G_1 with vertex set V_1 and edge set E_1 is a *subgraph* of a graph G_2 with vertex set V_2 and edge set E_2 if V_1 is a subset of V_2 and E_1 is a subset of E_2.

If m and n are positive integers with $m \leq n$, then K_n contains a subgraph that is isomorphic to K_m. In fact, any set of m of the vertices of K_n along with all possible edges between pairs of those m vertices forms such a subgraph. In turn, since any graph with m vertices is isomorphic to a subgraph of K_m, it must be the case that K_n contains subgraphs isomorphic to all graphs with n or fewer vertices.

Another type of graph that will be of interest to us is a *bipartite* graph.

Definition

Let G be a graph with vertex set V and edge set E. G is a *bipartite graph* if V is the union of two nonempty disjoint subsets V_1 and V_2 such that every edge of G has one endpoint in V_1 and the other endpoint in V_2.

If, in addition, E consists of all edges of the form $\{a, b\}$ where a belongs to V_1 and b belongs to V_2, then G is called a *complete bipartite graph*. If V_1 contains m vertices and V_2 contains n vertices, then this complete bipartite graph is denoted by $K_{m,n}$.

EXAMPLE 5.3

The graph shown on the left in Figure 5.3 is a bipartite graph but is not a complete bipartite graph. The graph shown on the right in the figure is the complete bipartite graph $K_{3,4}$.

• **Exercise 5.2**

(a) Copy the illustrations of the two bipartite graphs shown in Figure 5.3, and label the vertices of each of them. For each graph, determine two subsets of vertices that correspond to the sets V_1 and V_2 in the definition of bipartite graph.

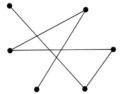 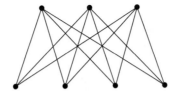

Figure 5.3 Two bipartite graphs. The second one is K$_{3,4}$.

(b) Using the labels for the vertices you used in part (a), write an isomorphism from the graph on the left in Figure 5.3 onto a subgraph of the graph shown on the right in that figure.

Some of the terminology of graph theory coincides with terminology already used in discussion of vertices and edges of tilings. This is illustrated in the final definition and example of this section.

> ### Definition
>
> The *valence* of a vertex *v* of a graph is the number of edges of the graph for which *v* is an endpoint.
>   A graph is called a *regular* graph if each of its vertices has the same valence.

EXAMPLE 5.4

Three regular graphs are illustrated in Figure 5.4. The first two are examples of *Platonic* graphs, those graphs consisting of the vertices and edges of the Platonic solids. Shown in the figure are the *tetrahedral graph* and the *octahedral graph*. The third regular graph shown is called the *Petersen graph*; it will be encountered later in this chapter.

• **Exercise 5.3**

(a) Which of the Platonic graphs are complete graphs? Explain your answer.
(b) Which of the Platonic graphs are bipartite?
(c) Which of the Platonic graphs are complete bipartite?

 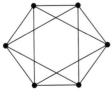 

Figure 5.4 Three regular graphs.

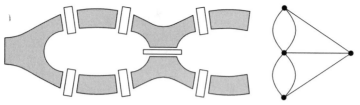

Figure 5.5 The bridges of Königsberg, represented by edges of a multi-graph.

5.2   PATHS IN GRAPHS

The common origin of the fields of topology and graph theory, as mentioned at the beginning of this chapter, was due to work Leonard Euler did in 1736. It was then that he presented a mathematical solution to a problem concerning the seven bridges of Königsberg, a town in East Prussia. The map of Figure 5.5 shows the seven bridges in the town and how they crossed the Pregel river, variously joining the two banks and two islands in the middle of the river. Apparently citizens of Königsberg had a recreational interest in trying to take a walk through town in such a way as to cross each bridge exactly once and return to the starting point. Euler's solution involved a "topological" representation of this situation as the "multi-graph" also shown in the figure. In this multi-graph, the four vertices represent the two banks and the two islands, and each of the seven edges represents a bridge connecting a bank and an island or the two islands. Note that this configuration differs from a graph, as we have defined it, only in that some pairs of vertices are joined by more than one edge.

In this section we will introduce some notions about paths in graphs that will allow us to solve a more general problem, and we will be able to draw Euler's conclusion about the bridge problem from our results. We will begin with a few simple definitions.

Definition

A *path* in a graph is a finite sequence of edges of the form $\{v_1, v_2\}$, $\{v_2, v_3\}, \ldots, \{v_{m-1}, v_m\}$ in which no edge is used more than once. We say that such a path *goes from* v_1 to v_m, we call v_1 the *initial vertex* of the path, and we call v_m the *final vertex* of the path.

A *closed path*, or *circuit*, in a graph is a path whose initial and final vertices are the same.

A graph is called *connected* if for each pair of distinct vertices there is a path from one of them to the other.

It is important for the reader to realize that the terminology introduced in this definition is not universal among all mathematicians. Indeed, what we have

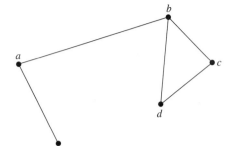

Figure 5.6 A connected graph. Edge $\{a, b\}$ is an isthmus.

defined as a *path* is called a "trail" by some authors of textbooks on graph theory, while "path" is used to denote a trail in which no vertex is visited more than once. In the treatment of this section, we do not need to make this distinction and, thus, opt to use the descriptive name *path* as given in the definition. This usage agrees with that of [**Wilson**].

The definition given above of connectedness of a graph agrees with our intuitive notion of connectedness of an object in the sense that a connected graph is not the union of two disjoint subgraphs. In fact, the word "components" is frequently used to refer to the pair-wise disjoint connected subgraphs of a graph, and so a connected graph is a graph with just one component. Note that a graph with just one vertex and no edges qualifies as a connected graph.

EXAMPLE 5.5

A connected graph is shown in Figure 5.6. In the graph, the edges $\{b, c\}$, $\{c, d\}$, and $\{d, b\}$ form a circuit, but there is no circuit that passes through both of the vertices a and b. Also, if the edge $\{a, b\}$ were to be deleted from the graph, it would become a disconnected graph. Such an edge—whose removal from a graph disconnects the graph—is called an *isthmus*.

EXAMPLE 5.6

Figure 5.7 shows a connected graph with no circuits. Such a graph is called a *tree*.

Euler's bridge problem involved determining, for his multi-graph, whether there existed a circuit that included all the edges. This idea leads to the following definition, which we present in the context of graphs.

> **Definition**
>
> An *Euler path* in a graph is a path that includes all the edges of the graph, and an *Euler circuit* is a closed Euler path.
> A graph is called *Eulerian* if it is connected and has an Euler circuit.

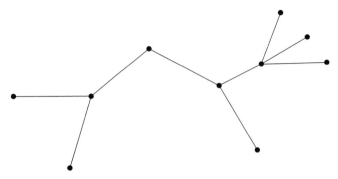

Figure 5.7 A connected graph with no circuits is a *tree*.

It is worth noting that a graph may have an Euler circuit and fail to be connected. An example is provided by a graph consisting of two components, an Eulerian graph and an isolated point.

EXAMPLE 5.7

Figure 5.8 shows three graphs. The graph labeled (*a*) is Eulerian; an Euler circuit is shown for that graph as the sequence of edges labeled 1 through 12. Graph (*b*) contains an Euler path (see Exercise 5.4), but it does not contain an Euler circuit, while graph (*c*) does not even contain an Euler path. We will address these assertions soon in our discussion.

• **Exercise 5.4** Find an Euler path for graph (*b*) of Figure 5.8.

We are now in a position to prove the main result of this section. It tells us precisely which connected graphs are Eulerian, and the proof will involve mathematical induction.

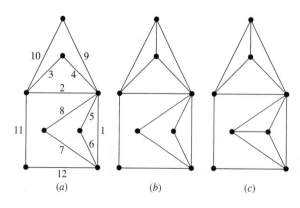

Figure 5.8 Of these graphs, only graph (*a*) is Eulerian.

> **Theorem**
>
> *Main Theorem on Eulerian Graphs*
>
> Let *G* be a connected graph. Then *G* is Eulerian if and only if every vertex of *G* has even valence.

There are two conditional statements to prove here, and one is easy and does not use mathematical induction. Assume that the connected graph *G* is Eulerian. Consider some specific Euler circuit of *G*. If *v* is a vertex of *G*, then each time that the circuit passes through *v*, it "enters" *v* along one edge and "leaves" *v* along a different edge, accounting for two distinct edges of *G* with *v* as an endpoint. Also, each time the circuit passes through *v*, a *different* pair of edges of *G* is encountered in this way. Since all edges of *G*—and consequently all edges of *G* with *v* as an endpoint—are included in the circuit, the valence of *v* must equal twice the number of times the circuit passes through *v*. So the valence of *v* is even, as desired.

To complete the proof, we will prove that if *G* is a connected graph in which every vertex has even valence, then *G* must have an Euler circuit. We will use the Principle of Complete Induction (which was introduced in Chapter 3), inducting on the number *n* of edges of *G*.

It is easy to see that the smallest number of edges for a connected graph in which every vertex has even valence is three. So the initial case of our induction argument is the case $n = 3$ in which the only graph to consider is K_3, the complete graph on three vertices. And this graph is clearly Eulerian.

So let *k* be a fixed, but arbitrary, positive integer greater than 3, and assume that all connected graphs with fewer than *k* edges and with all vertices of even valence possess Euler circuits. Now consider a connected graph *G* with exactly *k* edges and in which each vertex has even valence. We first claim that *G* has a circuit. To see this, start with any vertex v_1 of *G*. Select an edge e_1 from v_1 to some other vertex v_2. Then, since the valence of v_2 is even and, hence, greater than one, we can select an edge e_2 that is different from e_1 and connects v_2 to a vertex v_3. We can continue this process until, eventually (since *G* has only finitely many vertices), a previously selected vertex is repeated. The edges of the constructed path connecting that vertex to itself form a circuit in *G*.

So let such a circuit in *G* be denoted by *C*. If *C* contains all the edges of *G*, then *C* is an Euler circuit and the proof is complete. If *C* does not include all the edges of *G*, then the subgraph of *G* obtained by deleting all the edges of *C*, and any vertices that would remain as isolated vertices, is a subgraph with fewer than *k* edges. This subgraph may not be connected, but each of its "components" is connected. Moreover, any vertex in any one of those components *H* has even valence in *H*; indeed, a vertex of *H* is the endpoint of an even number of edges of *G*, and either all those edges remain in *H* or an even number of them are removed. So, by the induction hypothesis, any such component *H* has an Euler circuit. Now an Euler circuit of *G* can be constructed as follows: start in any of the above mentioned components at a vertex that also belongs to the circuit *C*; go around an Euler circuit in that component, returning to the original vertex; go along the circuit *C* until a vertex of another component is reached; go around an Euler circuit in that component; and continue this procedure until the original vertex is reached. So the desired result follows by mathematical induction.

EXAMPLE 5.8

It is now clear that neither of the graphs labeled (*b*) or (*c*) in Figure 5.8 is Eulerian. Graph (*b*) has two vertices of odd valence, and graph (*c*) has four vertices of odd valence. The assertion, made in Exercise 5.7, that graph (*c*) of Figure 5.8 does not admit even an Euler path follows from another result related to the Main Theorem on Eulerian Graphs: a connected graph admits an Euler path if and only if there are at most two vertices of odd valence. (See [**Wilson**].)

• **Exercise 5.5**   Which of the Platonic graphs are Eulerian? Explain your answer.

It is fairly easy to check that the above proof of the main theorem on Eulerian graphs can be modified to show that if a multi-graph has an Euler circuit, then each vertex must have even valence. Therefore, we may conclude—as did Euler—that the attempts by the residents of Königsberg were in vain!

Now that we have a condition that characterizes the Eulerian graphs, it is of interest to actually be able to find explicit examples of Euler circuits in such graphs. One straightforward method is described in the next theorem. We omit the proof, which may be found in [**Wilson**]. Recall that an *isthmus* in a connected graph is an edge whose removal disconnects the graph.

Theorem

Fleury's Algorithm

For any Eulerian graph *G*, the following construction always succeeds in producing an Euler circuit of *G*:

Arbitrarily select a starting vertex *v*. Starting at *v*, arbitrarily select *consecutive* edges of *G* according to the following rules:

(a) After selecting an edge *e*, the next edge must be chosen from the connected subgraph of *G* obtained by deleting *e*, all previously selected edges, and any vertices that become isolated when these edges are deleted. Call this subgraph the *allowable subgraph*.
(b) At each step, select an isthmus of the allowable subgraph only if it is unavoidable.

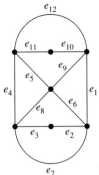

Figure 5.9 Fleury's algorithm produces this Euler circuit.

EXAMPLE 5.9

We may apply Fleury's algorithm to find an Euler circuit of the graph shown in Figure 5.9. For each of the 12 steps of the algorithm, the allowable subgraph (from which the ith consecutive edge e_i of the Euler circuit is selected) contains an available isthmus at steps $i = 3, 6, 7, 8, 9, 11,$ and 12, but selecting the isthmus is avoidable only in step $i = 6$.

• **Exercise 5.6** Confirm the steps of Fleury's algorithm applied in Example 5.9 to produce the Euler circuit shown in Figure 5.9. For each step, show the allowable subgraph from which the edge is selected.

Mathematicians have looked for other types of paths in connected graphs, and one type that is similar in definition to an Euler circuit is called a *Hamilton circuit*.

> **Definition**
>
> A *Hamilton path* in a graph is a path that passes through each vertex of the graph exactly once, and a *Hamilton circuit* is a closed Hamilton path.
> A graph is called *Hamiltonian* if it has a Hamilton circuit.

Note that a Hamiltonian graph is automatically connected.

EXAMPLE 5.10

The dodecahedral graph, the regular Platonic graph obtained from the vertices and edges of the regular dodecahedron, is a Hamiltonian graph. A Hamilton circuit on that graph is shown in Figure 5.10.

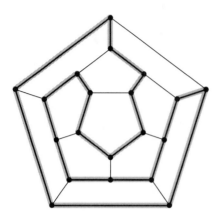

Figure 5.10 The dodecahedral graph is Hamiltonian.

• **Exercise 5.7** Show that the other four Platonic graphs are also Hamiltonian by drawing a Hamilton circuit on a diagram of each graph.

Mathematicians have not discovered any satisfactory theorem that characterizes the connected graphs that are Hamiltonian. In fact, the question of whether or not there is such a theorem is one of great interest in graph theory and has been related to ideas of computational complexity. The adjective "Hamiltonian" refers to the name of a 19th century Irish mathematician named William Hamilton who studied Hamilton circuits on the dodecahedral graph. Apparently, Hamilton was even involved in a project to market a mathematical puzzle consisting of a dodecahedron on which the 20 vertices were labeled with the names of famous cities. The player was to search for a path along which to tour all the cities, visiting each exactly once. This dodecahedron puzzle was perhaps the "Rubik's cube" of the 19th century!

5.3 EMBEDDING GRAPHS IN SURFACES

All of the results about graphs that we have discussed so far have been expressible in terms of the abstract definition of graph, without necessarily thinking about the edges of a graph as lines or arcs connecting the vertices, thought of as points in the plane or space. However, throughout all this discussion graph diagrams have been very useful—especially in clarifying some of the arguments involving paths. In this section, the property of graphs being representable by certain types of "spacial models" will be of primary interest.

When we have looked at examples of graph diagrams, it has often been necessary to let some of the lines or arcs representing edges cross each other in the plane at points other than those representing vertices. In certain cases, it was not clear that such intersections were absolutely necessary—perhaps some clever stretching or repositioning of some edges would have allowed us to avoid these extra crossings. This leads us to an important definition.

Definition

A graph is called a *planar graph* if it has a graph diagram in the plane such that no two arcs representing edges of the graph cross each other at any points other than those that represent common endpoint vertices. We will say that such a graph diagram does not have any *extra crossings*.

EXAMPLE 5.11

Graph diagrams for the complete graphs K_3, K_4, and K_5 were shown in Figure 5.2, and the diagram for K_3 given there certainly shows that K_3 is a planar graph. In Exercise 5.1 the reader was asked to attempt to draw diagrams for K_4 and K_5 without extra crossings. One such diagram for K_4 was shown in Figure 5.4 (as the tetrahedral graph), and this diagram confirms that K_4 is also a planar graph. The question of planarity of K_5 will be addressed later in this section.

EXAMPLE 5.12

Figure 5.11 shows two graph diagrams for the complete bipartite graph $K_{2,3}$. The diagram on the left, which is very natural to draw, has extra crossings. However, the diagram on the right, which contains no such extra crossings, illustrates that $K_{2,3}$ is a planar graph.

• **Exercise 5.8** Show that $K_{2,n}$ is planar for all $n \geq 1$ by providing a general diagram of the graph in the plane without any crossings. (Hint: Modify the second diagram shown in Figure 5.11 to obtain a diagram for $n = 4$ and 5. Then you should be able to draw a completely general picture!)

One of our objectives in this section is to discuss a necessary and sufficient condition for a connected graph to be planar. Our progress toward this objective will be aided by some topological ideas we have discussed in previous chapters.

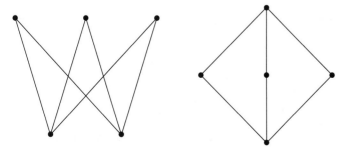

Figure 5.11 Two graph diagrams for $K_{2,3}$.

> *Definition*
>
> Let X be a compact surface, and let G be a graph. G is *embedded in X* if the vertices of G can be represented by points on X and the edges of G can be represented by arcs on X that connect the points representing their endpoint vertices in such a way that no two edge arcs meet except at the points representing common endpoint vertices. We will say that such a representation of G has *no extra crossings*.

Now any graph diagram can be drawn in the plane within a relatively "small" central region of the interior of a plane model for any compact surface. So if such a graph diagram for a graph G has no extra crossings, then clearly G can be embedded in any compact surface of our choosing. In particular, such a graph G can be embedded in the sphere S. Conversely, if a graph is embedded in S, then a representation of it without extra crossings on a plane model of S can be moved around on the plane model, if necessary, so as to lie entirely in the interior of the plane model. This allows us a first—and somewhat obvious—characterization of planar graphs.

> *Remark*
>
> A graph is planar if and only if it can be embedded in S.

EXAMPLE 5.13

Figure 5.12 shows three representations of the planar graph $K_{3,2}$ on the usual plane model for S. The first representation does *not* show that $K_{3,2}$ is embedded in S, but the second and third representations *do* show such embeddings. Note that, in the third representation, the edge arc that crosses the plane model boundary could be deformed on S so as to lie entirely in the interior of the plane model.

Some more examples will help to motivate the notion of embedding graphs in other compact surfaces.

EXAMPLE 5.14

By this point, the reader has probably tried—and failed—to produce a graph diagram for the complete graph K_5 without any extra crossings. So it seems likely that K_5 cannot be embedded in S. However, K_5 can be embedded in the torus T, as the representation on the usual plane model for T—shown in Figure 5.13—confirms.

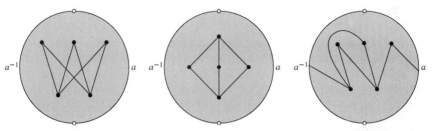

Figure 5.12 Representing $K_{3,2}$ on S. Only the second and third are embeddings.

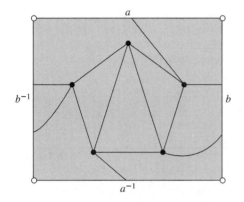

Figure 5.13 An embedding of K_5 in the torus.

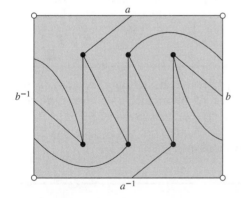

Figure 5.14 An embedding of $K_{3,3}$ in the torus.

EXAMPLE 5.15

Figure 5.14 shows a representation without extra crossings of the complete bipartite graph $K_{3,3}$ on the usual plane model of T. Thus, $K_{3,3}$ can be embedded in T.

It is a fact that any graph G can be embedded in some orientable compact surface. To see this, start with any graph diagram for a graph G. This diagram, drawn in the plane, can be thought of as a representation of G on the sphere S, but it may be a representation with extra crossings. Now for each such extra crossing of two edges, leave one edge on the sphere and slightly raise the other up off the sphere like a "bridge." Now a handle can be constructed and added to the sphere in such a way that the raised edge lies on the handle. The sphere with all necessary such constructed handles is homeomorphic to a multi-holed torus that provides an orientable compact surface in which G is embedded.

Of course, different graph diagrams for G may produce more or fewer handles and, therefore, multi-holed tori of different genera. In particular, there must be some orientable compact surface of least genus in which a given graph G can be embedded.

Theorem

For any graph G there is an orientable compact surface of smallest genus in which G can be embedded.

It is convenient to introduce some terminology associated with this last observation.

Definition

For a graph G, let the smallest genus of an orientable compact surface in which G can be embedded be denoted by $\gamma(G)$. $\gamma(G)$ is called the *genus* of the graph G.

If a graph G is embedded in mT, where $m = \gamma(G)$, then G is said to be *minimally embedded* in mT.

We should note that some authors prefer to consider the *characteristic* of a graph (i.e., the largest Euler characteristic of an orientable compact surface in which G can be embedded) in place of $\gamma(G)$. For example, see [**Firby**].

Since the smallest possible genus of orientable compact surface is 0, the genus of the sphere, the following observations are clear.

Remarks

(a) A graph G is planar if and only if $\gamma(G) = 0$.

(b) Any graph that is embedded in S is minimally embedded in S.

EXAMPLE 5.16

We have seen in previous examples that the graphs K_5 and $K_{3,3}$ can both be embedded in the torus T, which has genus 1. Therefore, $\gamma(K_5) \leq 1$ and $\gamma(K_{3,3}) \leq 1$. The remaining question of the possible planarity of these two graphs will be answered shortly.

To complete this section—and to be able to discuss the culminating ideas of this chapter—we will need to call up one of our most powerful tools: the Euler characteristic. We recall that, for an orientable compact surface X of genus g, we have $\chi(X) = 2 - 2g$. Also, we recall that if a tiling on a compact surface X has V vertices, E edges, and F faces, then $V - E + F = \chi(X)$. The important condition under which this latter formula holds is that of the definition of a tiling: that it be a covering of the surface *by polygonal faces* for which the interiors are "open cells"—in other words, sets homeomorphic to open planar disks.

Given a particular embedding of a graph in a compact surface, the vertex points and edge arcs indeed do determine regions of the surface, which we may call the *faces* of the embedding. Now, as Figure 5.15 illustrates, even for a graph embedded in the sphere, some of these faces may not turn out to be cells. Note that, for the graph illustrated, $V - E + F = 3$. However, it is a fact that if a *planar* graph G is *connected*, then the faces determined by an embedding of G in the sphere are cells. Therefore, we can draw the following conclusion.

Theorem

Let G be a connected planar graph and let a particular embedding of G in S determine V vertices, E edges, and F faces. Then $V - E + F = 2$.

EXAMPLE 5.17

Figure 5.16 shows two embeddings of connected graphs in S. The embedding shown on the left yields $V = 4$, $E = 3$, and $F = 1$. For the embedding of the graph shown on the right, $V = 4$, $E = 6$, and $F = 4$. In both cases, $V - E + F = 2$. The second graph is isomorphic to K_4.

Let us consider the possibility of planarity for the complete graph K_5. Now K_5 has 5 vertices and $\binom{5}{2} = 10$ edges. Let's suppose that in fact K_5 is planar. Then it can be embedded in the sphere, and the embedding determines $V = 5$ vertices, $E = 10$ edges, and some number F of faces satisfying $V - E + F = 2$.

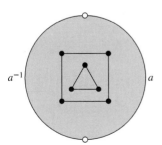

Figure 5.15 An embedding of a disconnected graph in the sphere. $V - E + F \neq 2$.

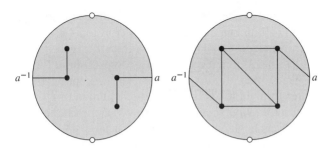

Figure 5.16 Two embeddings of connected graphs in S.

Thus, $2 = 5 - 10 + F$ and it follows that $F = 7$. Now since, in a graph, two vertices can be connected by at most one edge, none of the 7 faces will be 2-gon faces. So, using some counting methods familiar from Chapter 4, we have

$$3F = 3(F_3 + F_4 + F_5 + \ldots)$$
$$\leq 3F_3 + 4F_4 + 5F_5 + \ldots = 2E$$

But this certainly cannot be true since $3F = 3 \times 7 = 21$ while $2E = 2 \times 10 = 20$.

Therefore, K_5 is not a planar graph.

• **Exercise 5.9**   We have just shown that K_5 is not a planar graph. Do a similar proof to show that $K_{3,3}$ is not planar. (Hint: Think about the types of a–gon faces that an embedding of $K_{3,3}$ can determine.)

The last few results are summarized as a theorem.

Theorem

Neither the complete graph K_5 nor the complete bipartite graph $K_{3,3}$ is planar.

Recall that we noted, in Example 5.16, that both K_5 and $K_{3,3}$ can be embedded in the torus. So $\gamma(K_5) \leq 1$ and $\gamma(K_{3,3}) \leq 1$. Thus, it follows from the above theorem that $\gamma(K_5) = 1$ and $\gamma(K_{3,3}) = 1$.

It is interesting to note that there is a more general theorem that relates V, E, and F as determined by an embedding of a connected graph in orientable compact surfaces—possibly of positive genus. We state it here without proof.

Theorem

Let G be a connected graph and let a particular embedding of G in mT determine V vertices, E edges, and F faces.

If the embedding is a *minimal* embedding, then

(a) the faces determined by the embedding are all cells, and
(b) $V - E + F = 2 - 2m$.

EXAMPLE 5.18

Refer back to Figure 5.13, which showed an embedding of K_5 in the torus. For this embedding, $V = 5$, $E = 10$, and $F = 5$ so that $V - E + F = 0$, confirming the conclusion (b) of the theorem above. The reader can also easily check that all the faces are in fact cells, thereby confirming conclusion (a). The embedding must be minimal since K_5 is not planar.

EXAMPLE 5.19

A nonminimal embedding in mT may or may not determine faces that are cells and may or may not satisfy the formula $V - E + F = 2 - 2m$. Figure 5.17 shows three nonminimal embeddings of the planar graph K_4 in the torus T. Only the third embedding, shown on the right in the figure, determines faces that are cells and satisfies $V - E + F = 0$.

• **Exercise 5.10** For each of the three nonminimal embeddings of K_4 in the torus shown in Figure 5.17, determine V, E, and F and compute $V - E + F$.

The repeated appearances of the graphs K_5 and $K_{3,3}$ in the preceding discussion of planarity versus nonplanarity, are not just coincidental. These two graphs are intimately related to all nonplanar graphs by one of the greatest theorems common to graph theory and topology. It is universally known as *Kuratowski's Theorem* and is named for the mathematician who proved it in 1930. We will complete the chapter with a brief discussion of this important theorem.

Two connected graphs are called *homeomorphic* if they are isomorphic or if one can be obtained from the other by the "absorption" of some 2-valent vertices into their two incident edges so that they become one edge or by the insertion of some new 2-valent vertices in the middles of some edges breaking them into more edges. Two such homeomorphic graphs are shown in Figure 5.18 where the graph on the right is obtained from the graph on the left in this way.

Note that when two homeomorphic graphs are thought of as being embedded in some orientable surfaces, then the sets of points that their vertices and edges occupy on those surfaces will in fact be homeomorphic *objects* in the sense of homeomorphism with which we are already familiar. We now know enough to understand the statement of Kuratowski's Theorem.

Theorem

Kuratowski's Theorem

A connected graph is planar if and only if it contains no subgraph that is homeomorphic to either K_5 or $K_{3,3}$.

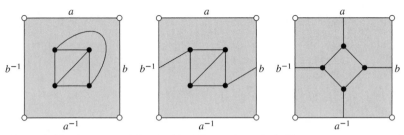

Figure 5.17 Three nonminimal embeddings of graphs in T.

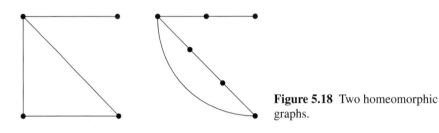

Figure 5.18 Two homeomorphic graphs.

EXAMPLE 5.20

The Petersen graph was shown earlier in this chapter in Figure 5.4 as an example of a regular graph: every vertex has valence 3. If one was to (correctly) conjecture that this graph is not planar, then it would seem natural to look for a "copy" of $K_{3,3}$ in the graph. However, there is no subgraph of the Petersen graph that is isomorphic to $K_{3,3}$ and certainly none that is isomorphic to K_5. However, some rearrangement of the graph diagram does reveal a subgraph that is homeomorphic to $K_{3,3}$. (See Exercise S-5.8 in this chapter's Supplementary Exercises, where some hints are provided!) So it follows from Kuratowski's Theorem that the Petersen graph is not planar.

5.4   CHAPTER SUMMARY

Graphs have been studied and applied to a variety of problems by mathematicians for the last three centuries, and the areas of graph theory and topology have shared many of the same ideas and supported each other throughout their history. In this chapter we have seen that graphs can be defined and investigated in an abstract way, independent of an interpretation as "line-and-dot" diagrams. However, the visual interpretation of a graph as being represented by a diagram not only gives us a way to picture the problems of graph theory but also motivates some of the interesting topological notions of graph theory. Investigating the ways of traveling along the arcs in the plane or in space that represent the edges of a graph leads to the significant problems of determining when graphs are Eulerian or Hamiltonian. Even today the problem of finding a suitable necessary and sufficient condition for a graph to be Hamiltonian remains unsettled, and this hard problem has influenced even the theory of computability and complexity. The simple observation that diagrams for some graphs cannot be drawn in the plane without extra crossings has led to the notions of planar graphs and embedding graphs in compact surfaces. The Euler characteristic proved useful in developing results concerning embeddings and ultimately in determining that the complete graph K_5 and the complete bipartite graph $K_{3,3}$ serve as obstacles to planarity for graphs, as expressed in Kuratowski's elegant theorem.

In the next—and final—chapter of this book, we will survey the mathematical theory of knots and links. Like graph theory, this area of topology has a rich history and continues today as an area of mathematics with many interesting open problems and applications.

5.5 SUPPLEMENTARY EXERCISES

Exercise S–5.1 For a graph G with vertex set V and edge set E, the *complement* of G is the graph $\bar{G}$ with vertex set V and edge set defined to be $\{\{a, b\}: a, b \in V, a \neq b,$ and $\{a, b\} \notin E\}$.

(a) Describe the complement of K_n.

(b) Describe the complement of $K_{m,n}$.

(c) Give an example of a graph with 4 vertices that is isomorphic to its complement. Can you also give such an example if you consider 5 vertices instead of 4? How about 6 vertices?

Exercise S–5.2 Suppose six people show up at a party. Show that either there are three people who all know each other or there are three people none of whom knows either of the other two. (Hint: Think of the six people as vertices of a graph in which two people are connected by an edge if and only if they know each other.)

Exercise S–5.3 Determine, with proof, the positive integers n for which

(a) the complete graph K_n is Eulerian.

(b) the complete graph K_n is Hamiltonian.

Exercise S–5.4 Determine, with proof, the pairs of positive integers m and n for which

(a) the complete bipartite graph $K_{m,n}$ is Eulerian.

(b) the complete bipartite graph $K_{m,n}$ is Hamiltonian.

Exercise S–5.5

(a) For each of the connected plane graph diagrams without extra crossings shown in Figure 5.19, determine whether the faces (regions of the plane determined by the diagram) can be colored using just two colors in such a way that each edge has two different colors alongside it. Note that the region "outside" of each diagram is to be considered a "face" to be colored. We will call such a plane graph diagram 2-*colorable*.

(b) Make a conjecture about the kind of connected plane graph diagrams without extra crossings that are 2-colorable.

Exercise S–5.6 A graph diagram, with extra crossings in the plane, for the octahedral graph was illustrated in Figure 5.4. Clearly this graph is planar since it represents the

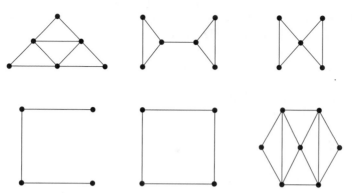

Figure 5.19 The graphs considered in Exercise S-5.5.

vertices and edges of a pattern on the sphere. Draw another diagram in the plane for the octahedral graph without any extra crossings.

Exercise S–5.7   Show that K_{11} cannot be embedded in the four-holed torus $4T$.

Exercise S–5.8

(a) Show that the Petersen graph is not planar by finding a subgraph that is homeomorphic to $K_{3,3}$. (Hint: Consider the two sets of vertices indicated by X's and O's in Figure 5.20.)

(b) Can you make a conjecture about the genus of the Petersen graph? Can you prove your conjecture?

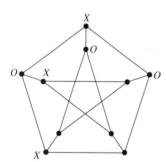

Figure 5.20 The diagram of the Petersen graph for Exercise S-5.8.

Chapter **6**

Knot Theory

K*nots*—objects that are homeomorphic to a circle—and their close relatives called *links*—disjoint unions of finitely many knots—are examples of compact 1-dimensional manifolds. These objects, and the way that they fit into 3-space, have been studied by mathematicians and scientists for well over a century, and that endeavor has even escalated during the last few decades. Of course, knots—in the context of tying loops of rope or thread—have been of interest to humankind since the earliest days of civilization. It is interesting that the earliest mathematical investigation and the recent resurgence of interest in knot theory share a common area of application, namely chemistry. In the late 1870s physicist Peter G. Tait was led to enumerate various knots based on their potential application to understanding Lord Kelvin's theory of the atom, a theory that related chemical properties of elements to knotted vortices of atoms. With recent work begun in the 1980s by topologists including DeWitt Sumners, Claus Ernst, and others, applications of knot theory to chemistry were finally realized, particularly in connection to research on recombinant DNA. For an account of such recent work, the reader is referred to [**Sumners**].

In this chapter we will introduce some of the basic mathematical ideas associated with the study of knots and links. When we consider whether two knots or links are *equivalent*, we will mean *isotopic in* 3-*space*. As pointed out in Chapter 1, we use the word "isotopic" to mean "ambiently isotopic" so that two links will be equivalent if and only if there is a continous deformation from one to the other that extends to all of 3-space, the "ambient" space of the deformation. As in Chapter 1, we will tacitly assume that all 3-space isotopies discussed are in fact such ambient isotopies. Therefore, the theory of knots and links in a sense is really the study of 3-space—the space in which continuous deformations from one knot or link to another are performed—rather than the study of knots and links themselves.

We will begin the chapter with a discussion of how the equivalence of two knots or links can be recognized in terms of their knot or link diagrams in the plane. A number of important and interesting examples will be introduced and used to motivate the need for some tools that will enable us to distinguish between inequivalent knots. Our look at these tools will culminate with investigation of some quite recently invented polynomial invariants of knots and links. Keeping with the tone of this book, our outlook will be intuitive, and we will not attempt to build up the rigorous ideas needed to prove all of the technical results discussed. The terminology and notation used throughout this chapter closely follows that of [**Livingston**].

6.1 KNOTS AND LINKS: THE BASICS

Imagine a piece of rope that we have tied into some knot in an apparently random way. Take the two loose ends and connect them—glue them together or sew them together—so that (ignoring the thickness of the rope) the resulting object can be thought of as a subset of 3-space homeomorphic to a circle. Now an interesting question to ask is whether or not the knotted rope can be "unknotted." Maybe, when tying the knot in a random fashion, we actually undid a previously knotted portion so that the rope can just be manipulated in space and fall into a simple circle! Now any of us who have experienced the frustration of trying to "unknot" a shoestring or a pulled out mangled cassette tape know that such a manipulation might be very complicated.

　　These ideas lead us to some first, basic definitions.

Definition

(a) A *knot* is a subset of 3-space that is homeomorphic to the unit circle.

(b) A *link* is a union of finitely many disjoint knots. The individual knots that make up a link are called its *components*. (So a knot is a link with just one component, i.e., a connected link.)

(c) Two links are called *equivalent* if they are isotopic in 3-space.

　　Note that two objects that are isotopic in 3-space must have the same number of components. Thus, any link that is equivalent to a knot must be a knot itself, and, more generally, if two links are equivalent, then they have the same number of components. Also note that saying components of a link are disjoint simply means that they do not share any common points. However, this use of the adjective "disjoint" does *not* imply that the components are "unlinked" in any way. We will be using some special terminology to refer to the simplest knots and links, which are in fact "unknotted" and "unlinked."

Definition

(a) Any knot that is equivalent to the unit circle $\{(x, y, 0) : x^2 + y^2 = 1\}$ is said to be *unknotted* and is called an *unknot.* We will use the symbol U to denote an unknot.

(b) Any link that is equivalent to $\{(x, y, i) : x^2 + y^2 = 1, i = 1, \ldots, n\}$ is called an *unlink of n components.* We will use U_n to denote an unlink of n components.

Thus, an unknot is just an unlink of one component (i.e., $U = U_1$). We will often be trying to decide whether a given knot (or link) is an unknot (or an unlink).

EXAMPLE 6.1

Figure 6.1 illustrates two knots we will see repeatedly in this chapter. These are the *left-handed trefoil knot,* which is shown on the left, and the *right-handed trefoil knot,* shown on the right. It seems intuitively clear that neither of these is an unknot. Looking beyond our intuition, however, it is somehow not entirely evident that there could not be *some*—possibly very complicated—continuous deformation that might "unknot" either of these knots. It might also be observed that the two knots shown look so similar that one might expect them to be equivalent to each other; the reader might try to find an appropriate continuous deformation. If these two knots are in fact inequivalent, then we would hope to be able to find some tool or method that will once and for all distinguish them.

EXAMPLE 6.2

Figure 6.2 shows three links, none of which *appear* to be unlinks. From left to right in the figure, these links are called the *Hopf link,* the *Whitehead link,* and the *Borromean link,* respectively.

• **Exercise 6.1** Explain why the Borromean link is not equivalent to either the Hopf link or the Whitehead link.

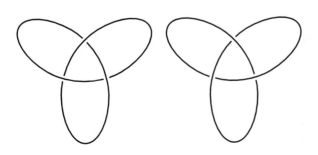

Figure 6.1 From left to right, the left-handed and right-handed trefoil knots.

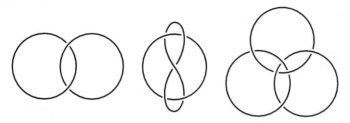

Figure 6.2 From left to right, the Hopf link, the Whitehead link, and the Borromean link.

• **Exercise 6.2** The link of three components shown in Figure 6.3 is formed from the Hopf link by adding on a new component as illustrated. Explain why this link of three components is not equivalent to the Borromean link. (Hint: Consider what happens when various components are removed from each of the links.)

Each of the knots or links shown in Figures 6.1, 6.2, and 6.3 was illustrated by displaying one of its *diagrams* in a plane of the page of the book. Of course, knots and links cannot be subsets of the plane unless they are unknots or unlinks. However, the *projection* of a knot or link on the plane does produce a subset of the plane that may include some apparent "crossing points." Any possible ambiguity about the nature of the actual knot or link near such a point can be resolved by using a gap to indicate a portion of the knot or link that passes under another portion. It should be fairly easy to believe that a diagram can be found that avoids any "bad" crossings, such as those shown in Figure 6.4. Now a given knot or link may have many very different looking diagrams, but if two knots or links can be shown to have the same diagram, then they are equivalent: from our intuitive point of view, they can be moved around in 3-space so that they both appear the same in a picture, namely, their common diagram. The converse of this statement is clearly true also.

Remark

Two knots or links are equivalent if and only if they can be shown to have the same diagram in the plane.

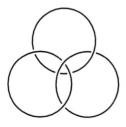

Figure 6.3 Another link of three components considered in Exercise 6.2.

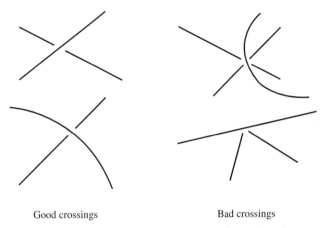

Good crossings   Bad crossings

Figure 6.4 Examples of some good and bad crossings for a diagram.

There is some useful notation that has traditionally been used to denote certain knots called *prime* knots. Prime knots are those that cannot be obtained as a certain "connected sum" of two other knots unless one of them happens to be an unknot. (A connected sum of two knots is formed by cutting each and connecting the two loose strands of the first knot to those of the second knot.) Prime knots are therefore considered "building blocks" of other knots, and in fact it has been shown that any given knot can be obtained as a connected sum of prime knots in a unique way. Now a given prime knot has many possible diagrams, but there must be a smallest number, say N, of crossings that occur in some of those diagrams. Over the last century, knot theorists have compiled prime knot tables in which prime knots are listed according to this minimum number N of crossings of their diagrams, and for each such N the corresponding prime knots are further enumerated with subscripts, as N_1, N_2, and so on—the order being based on tradition. Actually, not all prime knots have such designations since the traditional knot tables do not list both a prime knot and its reflection. The *reflection* of a knot is obtained by changing each crossing of a diagram of the knot so that the overcrossing strand becomes the undercrossing strand. If L denotes a knot or a link, then we will use the notation $\overline{L}$ to denote the reflection of L.

EXAMPLE 6.3

No diagram for either the right-handed or left-handed trefoil knot of Example 6.1, shown in Figure 6.1, can be drawn with fewer than three crossings, and both of these knots are prime knots. The left-handed trefoil knot is denoted by 3_1. Clearly the right-handed trefoil knot can be obtained as the reflection of the left-handed trefoil knot, so that it is denoted by $\overline{3_1}$. These are the only two prime knots for which the minimum number of crossings in any diagram is 3, and usually only 3_1 appears in a table of prime knots. Note that the question of equivalence of the left-handed and right-handed trefoil knots now becomes the question of whether the knot 3_1 is equivalent to its reflection $\overline{3_1}$.

Figure 6.5 The figure eight knot 4_1.

• **Exercise 6.3** Figure 6.5 shows a diagram for the prime knot 4_1, known as the *figure eight knot.*

(a) Draw a diagram of $\overline{4_1}$, the reflection of the figure eight knot.

(b) Using a piece of rope, actually construct the figure eight knot by knotting and connecting the ends of the rope so that when laid on a desktop it resembles the diagram shown in Figure 6.5. Then continuously deform your rope knot so that when laid out it resembles the diagram you drew in part (a). This shows that 4_1 is equivalent to $\overline{4_1}$.

Figure 6.6 provides a table of all prime knots with 7 or fewer crossings.

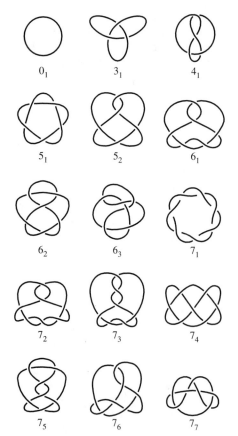

Figure 6.6 A short table of prime knots.

Reprinted by permission, Charles Livingston, *Knot Theory*, pp. 221–222, © 1993, The Mathematical Association of America.

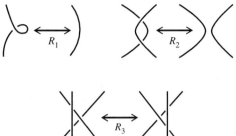

Figure 6.7 The Reidemeister moves.

The reader might have questioned the uniqueness of the reflection since its construction depends on a particular selected diagram for the original knot or link. This is just one incidence of many in which it would be convenient to have some more formal method for changing one diagram for a knot or link to another diagram for an equivalent knot or link. Fortunately, in 1927 a mathematician, Kurt Reidemeister, proved that all such changes of knot or link diagrams can be obtained by performing, repeatedly if necessary, three basic motions applied just to small portions of the diagrams near crossings, along with simple deformations in the plane, called *plane isotopies,* which do not change any of the crossings of diagrams. The three basic motions, now called *Reidemeister moves,* are denoted by R_1, R_2, and R_3 and are illustrated in Figure 6.7. Note that each of the moves illustrated in the figure actually represents several sorts of allowable changes in a diagram. First of all, each illustration shows how to change the arrangement on the left into the one on the right, and vice versa. Moreover, the illustrations for the moves could be "flipped over" to show allowable changes of slightly different crossings in some knots or links.

It is rather obvious that a finite sequence of Reidemeister moves will not change a knot or link represented by a diagram, as far as equivalence goes. It is the truth of the converse of this observation that is somewhat more stunning and useful for our purposes.

> ***Theorem***
>
> Two link diagrams in the plane represent equivalent links if and only if one can be converted into the other by a finite sequence of Reidemeister moves and plane isotopies.

We must keep in mind that, in our intuitive approach to defining knots and links and defining knot equivalence, we have necessarily avoided some technical and rigorous details needed to "formalize" these ideas. Indeed, the Reidemeister moves do give mathematicians a tool to treat the more formally defined idea of knot and link equivalence in an intuitive way—by looking at diagrams in the plane.

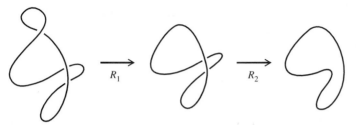

Figure 6.8 Reidemeister moves convert one diagram for an unknot to another.

EXAMPLE 6.4

Figure 6.8 illustrates how Reidemeister moves R_1 and R_2 can be used to change a knot diagram with three crossings into a diagram for an unknot with no crossings. So the original diagram is a diagram for an unknot, as the reader probably suspected in the first place!

EXAMPLE 6.5

Figure 6.9 shows how a certain knot diagram can be changed, using Reidemeister moves, into a knot diagram that we recognize from Figure 6.1 as representing the right-handed trefoil knot. Thus, the original diagram can be considered as a diagram for the right-handed trefoil knot.

• **Exercise 6.4**   Starting with the diagram for the figure eight knot 4_1 shown in Figure 6.5, show a sequence of Reidemeister moves that produces a diagram for $\overline{4_1}$ (i.e., the reflection of the diagram of Figure 6.5).

Reidemeister moves may also be used to modify diagrams for links.

EXAMPLE 6.6

In Figure 6.10, Reidemeister moves are used to show that the original link diagram illustrated is a diagram for the unlink of two components.

While Reidemeister moves provide a tool for demonstrating that two diagrams represent equivalent knots or links, they do not help at all in showing that two diagrams represent inequivalent knots or links. For example, in a technical mathematical sense, it is not clear from our discussion so far whether *all* knots are in fact equivalent to the unknot or whether *all* links with two components are equivalent to the unlink of two components. The remaining sections of the chapter will present some interesting methods for addressing such questions.

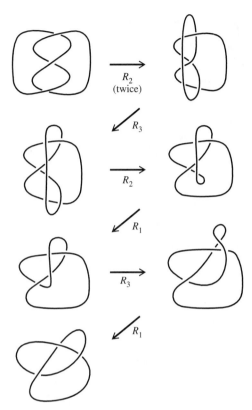

R_2
(twice)

R_3

R_2

R_1

R_3

R_1

Figure 6.9 Reidemeister moves convert one diagram for a right-handed trefoil knot into another.

6.2   ORIENTED LINKS AND LINKING NUMBER

Although we are interested in studying knots and links, as defined in the previous section, it is useful to add a little more structure to the objects we are studying, namely, the assignment of a "direction" to each component of a link.

Definition

We say that a link is *oriented* if a direction is assigned to each component of the link. Such an assignment of direction to a component will be indicated by one or more arrows drawn on the arcs of the link diagram in the plane.

Thus, a knot can be oriented in just two different ways while there are certainly more than two ways to orient a link of two or more components. Figure 6.11 shows three examples of oriented links of two components. Note that first and second links shown in the figure are, respectively, the Hopf link and the Whitehead link, which were shown previously (unoriented) in Figure 6.2. The

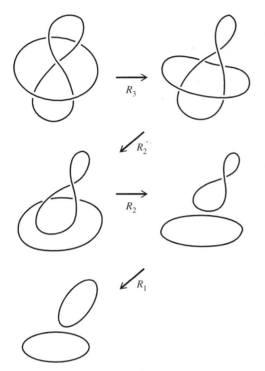

Figure 6.10 Reidemeister moves convert one diagram for an unlink of two components into another.

third link is called the King Solomon knot, even though it is not a "knot" at all! We will be primarily interested in oriented links of two components.

Of particular interest when considering a diagram for an oriented link of two components are the crossing points at which the two components meet. We will call a crossing point *right-handed* if an observer stationed on the overpassing arc and facing in the direction of that arc observes the underpassing arc's direction as from right to left. Otherwise, we will call the crossing point *left-handed*. These types of crossings, which are illustrated in Figure 6.12, are involved in the definition of the linking number of a link of two components.

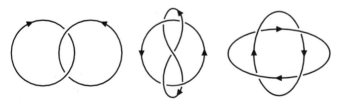

Figure 6.11 Three oriented links. The third one shown is called the King Solomon "knot."

Right-handed crossing   Left-handed crossing

Figure 6.12 Right-handed and left-handed crossings.

Definition

The *linking number* of an oriented link of two components is computed according to the following steps, applied to a diagram of the link:

(a) To each crossing point *at which the two components meet* assign the integer +1 if the crossing point is right-handed and the integer −1 if the crossing point is left-handed.

(b) The *linking number* of the oriented link is the sum of all the +1's and −1's divided by 2, unless there are no crossing points of the two components, in which case the *linking number* is 0.

This definition will need some discussion before it becomes clear that it is not ambiguous and before it becomes clear what service the linking number will provide us. But it will be helpful first to do a few examples of computing linking numbers.

EXAMPLE 6.7

Figure 6.13 shows two quite different diagrams of oriented unlinks of two components. As unoriented link diagrams, these were shown to be obtainable from each other in Example 6.6 and Figure 6.10. Note that in the first diagram there are no crossing points at all, so the linking number is just 0. In the second diagram there are five crossing points, but only four of them are crossing points at which the two components meet. In the figure, those four points are labeled with +1 or −1 appropriately, and so we see that the linking number for the second diagram is $\frac{(+1-1+1-1)}{2} = 0$, as it was for the first diagram also.

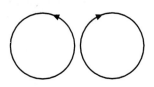 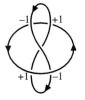

Figure 6.13 Two different oriented diagrams for U_2 yield the linking number 0.

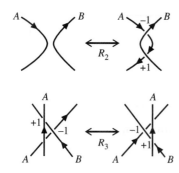

Figure 6.14 Applying R_2 or R_3 does not change linking number.

• **Exercise 6.5** Consider the three oriented links of two components shown in Figure 6.11. Copy each diagram, determine the appropriate $+1$ and -1 assignments for the crossings of the two components, and determine the linking number from each link diagram. Your results should yield linking numbers of 1, 0, and -2 for these particular oriented link diagrams of the Hopf link, the Whitehead link, and the King Solomon knot, respectively.

The results of Example 6.7 and Exercise 6.5 might suggest that the magnitude of the linking number is somehow involved with the number of times the two components of the link "link up" in a directed sense. Also, Example 6.7 suggests that it probably doesn't matter which diagram for an oriented link is used to compute the linking number. Reidemeister moves help to confirm this feeling.

It turns out that if one link diagram for an oriented link is changed into another diagram for an oriented link by any Reidemeister move, the linking number does not change. Figure 6.14 shows that this is true in special cases of moves R_2 and R_3. The various strands of the link are labeled A or B to denote the different components of the link. In each case, the net sum of the $+1$'s and -1's will certainly remain unchanged. The reader should check the other possible combinations of directions and component strands for moves R_2 and R_3. Of course, Reidemeister move R_1 involves strands from just one component of the link and need not even be checked.

By the above argument, we are justified in discussing the linking number of an oriented link of two components—we know that it will not depend on which diagram we use to do the computation. But another very important conclusion we can draw is that the absolute values of the linking numbers of two *equivalent* oriented links will be equal. After all, two equivalent oriented links can be considered to have the same diagram, possibly differing only in the directions of their components. Such a difference can account for either exactly the same sets of left-handed and right-handed crossings where components meet or an exchange of those types of crossings. So linking numbers are either equal or negatives of each other. What we have discovered, therefore, is an invariant of unoriented links of two components.

> **Theorem**
>
> If two equivalent (unoriented) links of two components are each oriented in any way, then the absolute values of their linking numbers will be equal.

This result is useful to us in its contrapositive form.

> **Theorem**
>
> If two (unoriented) links of two components are oriented in any way and the absolute values of their linking numbers are *not* equal, then the two links are *not* equivalent.

We are now able to say confidently that there are some links besides the unlink—a result that we have "known" intuitively all along. Any link of two components with nonzero linking number for some orientation cannot be equivalent to the unlink U_2. Since this intuitively plausible result required some fairly substantial work, we will designate it as a theorem.

> **Theorem**
>
> There exist links of two components that are not equivalent to the unlink U_2.

• **Exercise 6.6** For each of the following pairs of links of two components, determine whether absolute value of linking numbers can be used to determine whether the links are equivalent or inequivalent. If so, explain how. If not, explain why.

(a) The Hopf link and the Whitehead link.
(b) The Hopf link and the King Solomon knot.
(c) The Whitehead link and that King Solomon knot.
(d) The Whitehead link and the unlink U_2.

6.3   TRICOLORABILITY OF KNOTS AND LINKS

It is interesting to note that although we have now officially discovered many different links of two components, we have not yet officially proven that there are any knots that cannot be unknotted! The linking number does not apply to knots since a knot has just one component. In this section, we will learn about another invariant of links that also will help to distinguish knots.

We have already become comfortable with the idea of using a knot or link diagram in the plane to represent the actual knot or link being studied. Such a

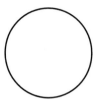 

Figure 6.15 Neither of these diagrams for the unknot U is tricolorable.

diagram consists of disjoint arc-shaped pieces that, when displayed together, give a picture of the actual knot or link. In this section, we will be concentrating on those little individual pieces that we will call the *arcs* of the diagram. In particular, we will consider a special way in which those arcs might be colored.

Definition

We will say that a knot or link is *tricolorable* if, given one of its diagrams and given a set of three colors, each arc of the diagram can be assigned one of the three colors in such a way that

(a) at least two of the colors are used, and

(b) if two different colors appear at any crossing, then so does the third color.

EXAMPLE 6.8

The simple circle diagram for the unknot U, shown on the left in Figure 6.15, cannot be tricolored since it has just one arc—the whole circle. The diagram shown on the right in that figure is also a diagram for the unknot (seen previously in Figure 6.8). This second diagram has just three arcs and three crossings. At one of the crossings, only two arcs appear, so those two arcs must be assigned the same color. Thus, using a second color for the third arc would leave two crossings at which two colors appear, and no third color appears. So this second diagram for the unknot is also not tricolorable.

EXAMPLE 6.9

A tricoloring of a diagram of the left-handed trefoil knot 3_1 is shown in Figure 6.16.

Figure 6.16 A tricolorable diagram for the left-handed trefoil knot 3_1.

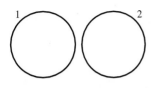 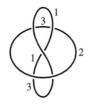

Figure 6.17 Both of these diagrams for U_2 are tricolorable.

EXAMPLE 6.10

The simple diagram, shown on the left in Figure 6.17, of the unlink of two components is clearly tricolorable. Of course, only two colors can be used, but a third is not needed since there are no crossings in this diagram. Another diagram for the same unlink is shown on the right in the figure, along with a tricoloring that does use all three colors. Note that at one of the crossings, just one color appears.

We would like to know that the property of tricolorability is an invariant for links. That is, we want to prove that if two links are equivalent and one of them is tricolorable, then so is the other. To see this, again we will look at Reidemeister moves. If a diagram for a link is tricolorable and is then changed into a new diagram for an equivalent link using just one application of a Reidemeister move, then we would like to show that the new diagram is also tricolorable. This can indeed be confirmed in all cases associated with each of the three Reidemeister moves. For example, the illustrations in Figure 6.18 show that application of move R_2 preserves tricolorability; the two possible coloring schemes of the arcs involved are shown in the figure.

• **Exercise 6.7**   Draw illustrations, similar to that shown in Figure 6.18, to demonstrate that Reidemeister moves R_1 and R_3 preserve tricolorability of link diagrams.

> **Theorem**
>
> If two links are equivalent and one of them is tricolorable, then so is the other.

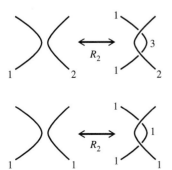

Figure 6.18 Reidemeister move R_2 preserves tricolorability.

As in the case of the absolute value of linking number for links of two components, it is again convenient to point out that the contrapositive form of this result will be useful in distinguishing knots and links.

> **Theorem**
>
> If a first link is tricolorable and a second link is not tricolorable, then the two links are *not* equivalent.

In Example 6.9 it was shown that one diagram for the left-handed trefoil knot 3_1 is tricolorable. Thus, we conclude that 3_1 is a tricolorable knot, unlike the unknot U for which no tricoloring exists. Thus, the knot 3_1 is not equivalent to the unknot U.

> **Theorem**
>
> There exist knots which are not equivalent to the unknot U.

• **Exercise 6.8**

(a) Determine whether the right-handed trefoil knot $\overline{3_1}$ is tricolorable.
(b) Does tricolorability distinguish $\overline{3_1}$ from the unknot U?
(c) Does tricolorability distinguish $\overline{3_1}$ from 3_1?

EXAMPLE 6.11

Figure 6.19 shows, once again, a diagram for the Whitehead link. Two of the arcs in this diagram are labeled a and b, and the crossing points are labeled P, Q, R, S, and T as shown. Suppose that we have a tricoloring of this diagram. Now if arcs a and b both have the same color, then since two arcs at P share the same color, the third arc at P must have

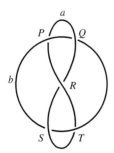

Figure 6.19 The Whitehead link is not tricolorable.

the same color as arcs *a* and *b*. Similarly, we may conclude that the third arc at *Q* has the same color as *a* and *b*. But then two arcs at each of the crossing points *S* and *T* share the same color, so the remaining arc must have the same color as arcs *a* and *b* too. This coloring of the arcs uses just one color, which contradicts the assumption that it is a tricoloring. On the other hand, suppose that arcs *a* and *b* have different colors; say that *a* is colored blue and *b* is colored red. Then the third arc at *P* must be colored a third color, say green, and so the third arc at *Q* must be colored green. Now one arc at *R* is colored blue and another arc at *R* is colored green, so that the third color, red, must be used for the third arc at *R*. This results in two red arcs and one green arc at *S*, which again contradicts the assumption that we have a tricoloring. Since arcs *a* and *b* either have the same color or different colors, we must conclude that this diagram is not tricolorable. Therefore, the Whitehead link is not tricolorable.

Since U_2, the unlink of two components, is tricolorable, we now know that the Whitehead link is not equivalent to U_2. Recall that this distinction was not determined by the linking number in the last section since both of these links have linking number 0.

- **Exercise 6.9**

(a) Diagrams for the Hopf link and the King Solomon Knot were shown in Figure 6.11. Ignoring the orientations shown there, use those diagrams to determine whether or not these links of two components are tricolorable.

(b) If two links of two components are distinguished by the absolute value of the linking number, then are they necessarily distinguished by tricolorability? Compare your answer to the conclusion drawn above about the Whitehead link and the unlink of two components.

6.4 POLYNOMIAL INVARIANTS OF KNOTS AND LINKS

One very important branch of topology is called *algebraic topology*. The techniques of algebraic topology often involve associating with each object (which may sometimes be a very abstract kind of object) some "algebraic quantity" that is known to be an "invariant" for the type of object being considered. Then, if two objects' associated values of this "quantity" are different, it follows that the two objects themselves must be different.

We have already seen several examples of this approach to distinguishing objects. In Chapter 4, we learned how to associate with each compact surface *X* a number $\chi(X)$, the Euler characteristic of *X*. And, indeed, it turns out that if two compact surfaces are homeomorphic, then they have the same Euler characteristic; so compact surfaces with different Euler characteristics are indeed different (nonhomeomorphic). We say that the Euler characteristic is an *invariant* for compact surfaces. Another example of an algebraic invariant was introduced in Section 2 of this chapter. There we learned that the absolute value of the linking number of a link of two components is an invariant of such links in the sense that if two links of two components are equivalent, then the absolute values of their linking numbers must be equal. Thus, two links of two components for which the absolute values of linking numbers differ are not equivalent links.

One of the great successes of algebraic topology has involved the introduction of very powerful algebraic invariants that are more complicated than simple numbers. In this section, we will study two algebraic invariants of links (and, hence, knots) that assign to each link a *polynomial* in some variable. As is often the case with algebraic invariants considered in advanced algebraic topology, the assignment process may seem at first somewhat ad hoc and mysterious. But the true justification of considering such invariants is the fact that they can prove to be useful for telling apart objects (knots or links, for us) that we have not been able to distinguish previously by other means.

In 1928 a mathematician named J. W. Alexander proved that a polynomial could be associated with each knot so that equivalent knots are assigned the same polynomial, and this *polynomial invariant* proved to be very good at distinguishing knots. We will develop the ideas of this invariant in a way presented in the 1970s by J. H. Conway, and the polynomial considered here will be called the Conway polynomial.

Consider a diagram of an *oriented* link L in the plane and focus attention on some selected crossing. This crossing will be either a left-handed crossing or a right-handed crossing. (Refer back to Section 2 of this chapter for the definitions of these crossing types.) If the over-crossing strand and under-crossing strand are moved through each other, the type of crossing (and the link diagram itself) is changed. We will let L_+ denote the link obtained from L when the selected crossing is arranged so as to be right-handed, and we will use L_- to denote the link obtained when the selected crossing arrangement is left-handed. Note that one of these links L_+ or L_- is in fact the oriented link L. Also associated with L is a third link, denoted by L_s, which is obtained from L (or from L_+ or from L_-) by "smoothing out" the selected crossing; the crossing is actually removed and the four "loose" strands are reattached in the one possible way to preserve the directions of the oriented components. The general appearance of the links L_+, L_-, and L_s near the selected crossing is shown in Figure 6.20.

EXAMPLE 6.12

A diagram for the unknot, oriented in a certain way and containing one right-handed crossing, is shown as L_+ in Figure 6.21. The corresponding diagrams for L_- and L_s are also shown in the figure. Note that L_- is also a diagram for an oriented unknot, while L_s is a diagram for an oriented unlink of two components.

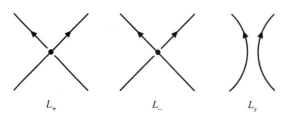

L_+ L_- L_s

Figure 6.20 How L_+, L_-, and L_s appear near the selected crossing.

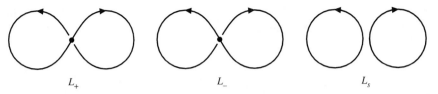

Figure 6.21 L_+, L_-, and L_s for the link of Example 6.12.

We will now present the definition of the Conway polynomial of an oriented link. For an oriented link L, the variable of the polynomial will be taken to be the letter z and the Conway polynomial itself will be denoted by $\nabla_L(z)$. The definition is "implicit" in the sense that the polynomials associated with three general links L_+, L_-, and L_s are related by an equation. Then, given the polynomial associated with the unknot, specific polynomials for other more complicated oriented links may be explicitly determined from this relation in a "recursive" manner.

> ### Definition
>
> **(a)** The Conway polynomial for the unknot U is defined by $\nabla_U(z) = 1$.
> **(b)** For the links L_+, L_-, and L_s, related as described above, the Conway polynomials for L_+, L_-, and L_s satisfy $\nabla_{L_+}(z) - \nabla_{L_-}(z) = -z\nabla_{L_s}(z)$.

Since this definition does not directly tell us how to immediately compute the polynomial $\nabla_L(z)$ for a given oriented link L, there is some question of whether or not such a polynomial can always be found. It is the case that such polynomials always exist, and this result was proven by Alexander and, for the modified version of Alexander's polynomial we are using, by Conway.

EXAMPLE 6.13

Refer back to Figure 6.21 where both L_+ and L_- are the unknot. (Note that, up to equivalence, there is only one way to orient an unknot.) In this case, $\nabla_{L_+}(z) = \nabla_{L_-}(z) = 1$. So, the equation $\nabla_{L_+}(z) - \nabla_{L_-}(z) = -z\nabla_{L_s}(z)$ yields $1 - 1 = -z\nabla_{L_s}(z)$, whence $\nabla_{L_s}(z) = 0$. In this case, L_s is an oriented unlink of two components. Since, up to equivalence, there is only one way to orient an unlink of two components, we conclude that such a link has Conway polynomial 0.

It should be clear to the reader that some of the language used in the previous example relies on a big assumption, namely that the Conway polynomial depends only on the oriented link and not on the particular diagram and crossing used to do the computation. A proof that this is true would take the form of verifying that changing a link via any Reidemeister move will produce a link with the same polynomial. We will not undertake this proof, but instead just highlight these important facts as a theorem.

Theorem

The Conway polynomial of an oriented link remains unchanged if the diagram used to compute its polynomial is modified by any Reidemeister move.

Therefore, the Conway polynomial of an oriented link depends only on the oriented link and not on the particular diagram used to compute the polynomial.

Two oriented links may be equivalent as links, without regard to their orientations. However, "equivalence *as oriented links*" requires more.

Definition

Two oriented links are *equivalent as oriented links* if there is a continuous deformation in 3-space from one to the other that preserves the direction of the components.

Thus, two oriented links that are equivalent as oriented links are automatically equivalent links since they are isotopic in 3-space. Also, two oriented links that are equivalent as oriented links will possess a common *oriented* link diagram. With this comment, the previous theorem immediately leads to the fact that the Conway polynomial is indeed an invariant of oriented links.

Theorem

If two oriented links are *equivalent as oriented links,* then they have the same Conway polynomial.

Or, in contrapositive form:

If two oriented links have different Conway polynomials, then they are not *equivalent as oriented links.*

EXAMPLE 6.14

There are two distinct ways to orient the usual diagram of the Hopf link. Figure 6.22 shows these two diagrams. The diagram on the left in the figure has both of its crossing points left-handed, and we will refer to the associated oriented link as the *left-handed Hopf link*. In the diagram on the right the direction of one component is changed, and

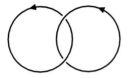 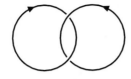

Figure 6.22 From left to right, the left-handed and right-handed Hopf links.

both crossings are right-handed. The associated oriented link is the *right-handed Hopf link.*

EXAMPLE 6.15

Let us compute the Conway polynomial of the left-handed Hopf link. In Figure 6.23, it appears as L_-. In this case, L_+ is the unlink of two components and L_s is the unknot. Thus, $\nabla_{L_+}(z) - \nabla_{L_-}(z) = -z\nabla_{L_s}(z)$ yields $0 - \nabla_{L_-}(z) = -z \cdot 1$ and we conclude that $\nabla_{L_-}(z) = z$.

• **Exercise 6.10**   Use the diagram of the right-handed Hopf link shown in Figure 6.22 to verify that its Conway polynomial is $-z$. Are the left-handed and right-handed Hopf links equivalent as oriented links?

When we restrict our attention to Conway polynomials for *knots,* the question of which way the knot has been oriented becomes moot. Each crossing of a knot diagram is either left-handed or right-handed *without regard to how the diagram is oriented.* It follows that we can talk about the Conway polynomial as an invariant of *unoriented* knots.

Theorem

If two knots have different Conway polynomials, then they are not equivalent.

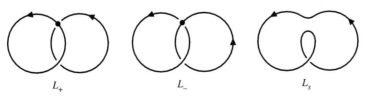

Figure 6.23 The left-handed Hopf link appears here as L_-.

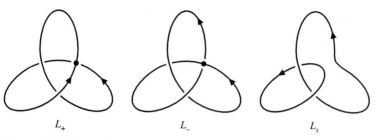

$$L_+ \qquad\qquad L_- \qquad\qquad L_s$$

Figure 6.24 Here L_- is the left-handed trefoil knot 3_1.

EXAMPLE 6.16

We can use some previous results to compute the Conway polynomial of the left-handed trefoil knot 3_1. Figure 6.24 shows a related sequence L_+, L_-, and L_s in which 3_1 appears as L_-, L_+ is the unknot, and L_s is the left-handed Hopf link. Thus, $\nabla_{L_+}(z) - \nabla_{L_-}(z) = -z\nabla_{L_s}(z)$ yields $1 - \nabla_{L_-}(z) = -z \cdot z$, whence $\nabla_{L_-}(z) = z^2 + 1$.

EXAMPLE 6.17

With the help of Figure 6.25, we can do a computation, similar to that of the previous example, of the Conway polynomial of the right-handed trefoil knot $\overline{3_1}$. In that figure, L_+ is the right-handed trefoil knot $\overline{3_1}$, L_- is the unknot, and L_s is the right-handed Hopf link. Recall, from Exercise 6.10, that $\nabla_{L_s}(z) = -z$. So this time it follows from $\nabla_{L_+}(z) - \nabla_{L_-}(z) = -z\nabla_{L_s}(z)$ that $\nabla_{L_+}(z) - 1 = -z \cdot (-z) = z^2$. Thus, the Conway polynomial of the right-handed trefoil knot is $\nabla_{L_+}(z) = z^2 + 1$, the same polynomial obtained as the Conway polynomial of the left-handed trefoil knot in the previous example.

So, even with the help of the Conway polynomial as an invariant of knots, we still have failed to determine whether the two versions of the trefoil knot are equivalent! Thus, we remain motivated to seek yet another invariant that perhaps will be strong enough to distinguish the reflected trefoil knots.

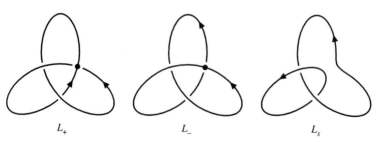

$$L_+ \qquad\qquad L_- \qquad\qquad L_s$$

Figure 6.25 Here L_+ is the right-handed trefoil knot $\overline{3_1}$.

In the late 1980s L. Kauffman introduced a polynomial invariant of knots that has proven to be very useful in the advanced study of knot theory. We will introduce the definition of the Kauffman bracket polynomial for oriented links and then apply it to distinguish a few knots. Of course, as the reader should expect, the fact that this polynomial is an invariant involves checking that it does not change under application of Reidemeister moves to link diagrams; we will not attempt the details of this verification here.

There is one technical point to discuss first. Kauffman's polynomial is not a polynomial in the sense we usually mean in elementary algebra. Instead it is a Laurent polynomial. A *Laurent polynomial* is like a usual polynomial except that it allows negative integer powers of the polynomial variable as well as nonnegative integer powers. So every usual polynomial is a Laurent polynomial, but some Laurent polynomials are not usual polynomials. For example, if the polynomial variable is the letter t, then $t^{-12} + 3t^{-4} + t^8$ is a Laurent polynomial but not a usual polynomial.

Let us begin with a diagram for an *unoriented link*. With just a slight rotation, each crossing can be arranged so that the over-crossing strand goes from "southwest" to "northeast." Starting from this arrangement at a crossing, there are two ways of "smoothing" the crossing so that the crossing is removed and pairs of strands are connected. One way to do this involves connecting the "northwest" strand to the "southwest" strand and the "northeast" strand to the "southeast" strand—a *type A* smoothing. The alternative is to connect the "northwest" strand to the "northeast" strand and the "southwest" strand to the "southeast" strand—a *type B* smoothing. The general situation for these two smoothings of such a crossing is shown in Figure 6.26.

Let us first describe the idea of computing the *Kauffman bracket* of an unoriented link. The Kauffman bracket will be a Laurent polynomial in the variable t, but it will just be an algebraic factor of the polynomial that will be of ultimate interest. If we start with a diagram for a link and situate its crossings as mentioned above, then we can apply one of the two smoothing types *at each crossing*. (If there are n crossings, then there are 2^n ways to do this.) This will produce a modified link diagram called a *state* of the original diagram in which there will be *no crossings*, and such a modified diagram will consist entirely of nonintersecting closed curves (i.e., "circles"). The Kauffman bracket is defined in terms of the number of such circles and the numbers of type A and type B smoothings for all possible states of the original diagram.

Appropriately situated crossing

Type *A* smoothing

Type *B* smoothing

Figure 6.26 Type *A* and Type *B* smoothings of a crossing.

> ***Definition***
>
> For a diagram D of an unoriented link L with crossings arranged as described above, a *state* of D is the diagram that results from applying either a type A or type B smoothing at each crossing of D.
>
> For each state S of D, define $\langle D|S \rangle = t^{a-b}$ where a is the number of type A smoothings of the state and b is the number of type B smoothings of the state.
>
> Define the *Kauffman bracket* $\langle D \rangle$ of the diagram D to be the sum of all the terms of the form $\langle D|S \rangle (-t^{-2} - t^2)^{|S|-1}$, one such term for each state S of D, where $|S|$ denotes the number of circles of S. Written symbolically, $\langle D \rangle = \Sigma_S \langle D|S \rangle (-t^{-2} - t^2)^{|S|-1}$.

To illustrate that this computation can be easier, at least in some cases, than this rather imposing equation might suggest, let us look at some simple examples.

EXAMPLE 6.18

We will compute the Kauffman bracket for the diagram D of the unknot U consisting of a simple circle in the plane. Since there are no crossings in D there is just $2^0 = 1$ state, namely $S = D$. For this state S, we have $a = b = 0$ and $|S| = 1$. Thus, for that one state S, $\langle D|S \rangle = t^{a-b} = t^0 = 1$, and so the Kauffman bracket of D is $\langle D \rangle = \langle D|S \rangle (-t^{-2} - t^2)^{|S|-1} = 1(-t^{-2} - t^2)^{1-1} = 1(-t^{-2} - t^2)^0 = 1$.

EXAMPLE 6.19

Let us compute the Kauffman bracket for a different diagram D of the unknot shown in Figure 6.27. The one crossing of D is shown situated in the appropriate way to compute the Kauffman bracket, and the two states of D are shown in the figure labeled as S_1 and S_2. Note that for S_1, $a = 1$, $b = 0$, and $|S_1| = 2$ while, for S_2, $a = 0$, $b = 1$, and $|S_2| = 1$. Thus, $\langle D|S_1 \rangle = t^{1-0} = t$, and $\langle D|S_2 \rangle = t^{0-1} = t^{-1}$ so that the Kauffman bracket for D is

$$
\begin{aligned}
\langle D \rangle &= \langle D|S_1 \rangle (-t^{-2} - t^2)^{|S_1|-1} + \langle D|S_2 \rangle (-t^{-2} - t^2)^{|S_2|-1} \\
&= t(-t^{-2} - t^2)^{2-1} + t^{-1}(-t^{-2} - t^2)^{1-1} \\
&= t(-t^{-2} - t^2)^1 + t^{-1}(-t^{-2} - t^2)^0 \\
&= -t^{-1} - t^3 + t^{-1} \\
&= -t^3
\end{aligned}
$$

It is clear from the last two examples that the Kauffman bracket is *not* an invariant of unoriented links since the diagrams of Examples 6.18 and 6.19 can be obtained from each other by application of Reidemeister move R_1. However, it is the case that the Kauffman bracket *is* preserved by Reidemeister moves R_2 and R_3.

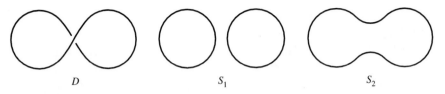

Figure 6.27 A diagram for the unknot and its two states.

• **Exercise 6.11** Find the Kauffman bracket of the diagram of the Hopf link shown in Figure 6.28. Note that the two crossings are appropriately situated for the computation and that there are four states to consider.

A slight modification of the Kauffman bracket produces a Laurent polynomial that *is* an invariant of *oriented* links.

Definition

Let D be a diagram of an oriented link L, let r denote the number of right-handed crossings of D, and let l denote the number of left-handed crossings of D.

The *Kauffman bracket polynomial* of D is defined to be $(-t)^{-3(r-l)}\langle D \rangle$ where $\langle D \rangle$ is the Kauffman bracket of D.

EXAMPLE 6.20

If we let D be the simple circle diagram for the unknot considered in Example 6.18, then $\langle D \rangle = 1$. Since that diagram has no crossings, $r = l = 0$ for either of the two possible orientations of D. Thus, the Kauffman bracket polynomial for D is $(-t)^{-3(r-l)}\langle D \rangle = (-t)^0 \times 1 = 1$. Now if instead we let D denote the diagram for the unknot considered in Example 6.19 and supply either of the possible orientations, we will have $r = 1, l = 0$, and $\langle D \rangle = -t^3$. Thus, in this case, the Kauffman bracket polynomial of D is $(-t)^{-3(r-l)}\langle D \rangle = (-t)^{-3(1-0)}(-t^3) = (-t^{-3})(-t^3) = 1$. The Kauffman bracket polynomial for each of these oriented diagrams of the unknot equals 1.

The fact that the Kauffman bracket polynomial is an invariant is summarized in the following theorem.

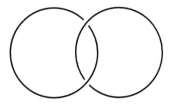

Figure 6.28 The diagram for the Hopf link considered in Exercise 6.11.

Theorem

Let L be an oriented link. Any two diagrams for L will yield the same Kauffman bracket polynomial. In other words, we are justified in speaking of the *Kauffman bracket polynomial of L*. If we denote this polynomial by $F_L(t)$, then $F_L(t) = (-t)^{-3(r-l)} \langle D \rangle$ where D is any diagram for L, $\langle D \rangle$ is the Kauffman bracket of D, and r and l are the numbers of right- and left-handed crossings of D, respectively.

Moreover, if L is a knot, then $F_L(t)$ does not depend on the selected orientation of L, so that $F_L(t)$ is an invariant of unoriented knots.

• **Exercise 6.12**   Orient the diagram for the Hopf link shown in Figure 6.28 by supplying a direction to each of its components so that both crossings are right-handed, thus producing the right-handed Hopf link. Then use the result of Exercise 6.11 to find the Kauffman bracket polynomial for the right-handed Hopf link.

EXAMPLE 6.21

Let us consider the diagram, with three crossings, of the right-handed trefoil knot 3_1 shown in Figure 6.29. In that figure, below the original knot, are shown the $2^3 = 8$ states, labeled S_1 through S_8.

The following table shows all the appropriate information about these eight states necessary for the computation of the Kauffman bracket of this knot diagram, which turns out to be $\langle D \rangle = (-t - t^5) + 3t + 3(-t^{-3} - t) + (t^{-7} + 2t^{-3} + t) = t^{-7} - t^{-3} - t^5$.

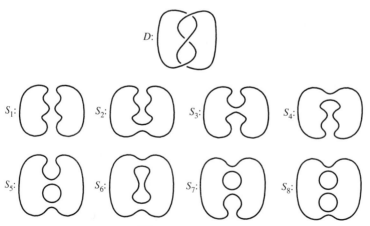

Figure 6.29 A diagram for $\overline{3}_1$ and its eight states.

	a	b	$|S|$	$\langle D|S\rangle = t^{a-b}$	$\langle D|S\rangle(-t^{-2}-t^2)^{|S|-1}$
S_1	3	0	2	t^3	$-t - t^5$
S_2	2	1	1	t	t
S_3	2	1	1	t	t
S_4	2	1	1	t	t
S_5	1	2	2	t^{-1}	$-t^{-3} - t$
S_6	1	2	2	t^{-1}	$-t^{-3} - t$
S_7	1	2	2	t^{-1}	$-t^{-3} - t$
S_8	0	3	3	t^{-3}	$t^{-7} + 2t^{-3} + t$

Now, when supplied with either of the possible orientations, all three crossings are right-handed yielding $r = 3$ and $l = 0$. Thus, the Kauffman bracket polynomial of $\overline{3_1}$ is

$$F_{\overline{3_1}}(t) = (-t)^{-3(r-l)}\langle D\rangle$$
$$= (-t)^{-9}(t^{-7} - t^{-3} - t^5)$$
$$= -t^{-16} + t^{-12} + t^{-4}$$

• **Exercise 6.13**   Use the diagram shown in Figure 6.30 to show that the Kauffman bracket polynomial for the left-handed trefoil knot 3_1 is $F_{3_1}(t) = -t^{16} + t^{12} + t^4$. Note that the crossings have been situated appropriately for the computation of the Kauffman bracket.

Given the results of Example 6.21 and Exercise 6.13 and the fact that the Kauffman bracket polynomial is an invariant of unoriented knots, we finally see that the left-handed and right-handed trefoil knots are inequivalent knots.

6.5   CHAPTER SUMMARY

In the final chapter of this book we have discussed a rich area of mathematics that has its origin in the nineteenth century and is also of much current interest, not only to mathematicians, but also to scientists who see potential for using the ideas of knot theory to tackle significant problems of science. We have learned

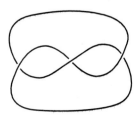

Figure 6.30   The diagram for 3_1 for Exercise 6.13.

that equivalence of knots or links involves the intuitively understood idea of compact curves being isotopic in 3-space. Moreover, such continuous deformations of knots and links can be interpreted in terms of Reidemeister moves, allowing for an easy description of such equivalences.

To address the idea of using certain invariants to establish the nonequivalence of certain knots or links, we considered just a few of many tools that topologists have developed over the years. To distinguish some links of two components, we introduced the idea of linking number, and tricolorability served to distinguish between some knots as well as some links. Knot and link polynomials proved to be somewhat stronger tools for distinguishing knots and links, and they also served as an important example of algebraic invariants of topological objects. These polynomials, which have been introduced fairly recently, have been valuable to topologists in solving previously unsolved problems of knot theory, and their use has also generated new and interesting questions, many of which remain unanswered and await solution by mathematicians of the future.

The survey of knot theory presented in this chapter—like the surveys of the topological theories of manifolds, compact surfaces, tilings, and graphs provided in the previous chapters of the book—hopefully has inspired readers to learn more about the subject of topology and to be a part of the group of mathematical enthusiasts and researchers of the future.

6.6   SUPPLEMENTARY EXERCISES

Exercise S-6.1   What can you say about a knot that has a diagram with exactly two crossings?

Exercise S-6.2   A knot diagram is called *alternating* if the overpasses and underpasses alternate as the diagram is traversed, and an *alternating knot* is one that has an alternating diagram. For example, a trefoil knot is an alternating knot.

(a) Draw an alternating knot diagram for the unknot with exactly four crossings.

(b) Draw an alternating knot diagram with exactly four crossings that can be redrawn as an equivalent diagram with just three crossings but not as an equivalent diagram with fewer than three crossings.

(c) Draw an alternating knot diagram with exactly four crossings that cannot be redrawn as an equivalent diagram with fewer than four crossings.

Exercise S-6.3   The knot whose diagram is shown in Figure 6.31 is an unknot. Use Reidemeister moves to change it to a simple circle diagram for the unknot.

Figure 6.31 The unknot diagram for Exercise S-6.3.

Exercise S-6.4   A diagram for an oriented link L of two components appears in Figure 6.32.

(a) Find the linking number of L.

(b) Ignore the orientation of L. Is L tricolorable? If so, illustrate a tricoloring of the arcs of the diagram. If not, carefully explain why not.

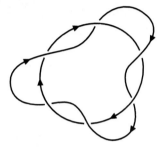

Figure 6.32 The link diagram for Exercise S-6.4.

Exercise S-6.5   Find an example of a knot, other than the unknot, that is not tricolorable and prove that it is not tricolorable.

Exercise S-6.6   Find the Conway polynomial of the oriented Whitehead link for which a diagram is shown in Figure 6.33.

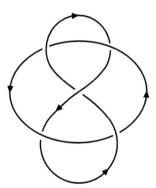

Figure 6.33 The oriented Whitehead link diagram for Exercise S-6.6.

Exercise S-6.7   Find the Conway polynomial for the figure eight knot 4_1.

Exercise S-6.8   Show that the Kauffman bracket polynomial of the unknot diagram shown in Figure 6.34 with three crossings is 1.

Figure 6.34 The unknot diagram for Exercise S-6.8.

Appendix

REFERENCES AND SUGGESTIONS FOR FURTHER STUDY

For the reader who wishes to learn more about topology, we list here a selection of available resources with brief descriptions. Included here are the sources referred to in the text.

- Edwin Abbott, *Flatland* (Dover, 1952).

  A short treatise from the late nineteenth century that relates the story of a 2-dimensional being, A. Square; a must to read for anyone interested in higher dimensional analysis; also includes commentary on societal issues.

- Colin C. Adams, *The Knot Book* (W. H. Freeman, 1994).

  A good introduction to knot theory; very readable, with modest prerequisites.

- H. R. Brahana, "Systems of circuits on two-dimensional manifolds," *Annals of Mathematics* **23** (1922), pp. 144–168.

  The historical article in which Brahana presents his original approach to proving the classification theorem for compact surfaces. Our approach to the proof is based on this "method of Brahana."

- H. S. M. Coxeter, *Introduction to Geometry* (John Wiley & Sons, 1961).

  A classic among introductory geometry books; includes some material on topology that addresses global regularity of tilings.

- Allan L. Edmonds, John H. Ewing, and Ravi S. Kulkarni, "Regular tessellations of surfaces and $(p, q, 2)$-triangle groups," *Annals of Mathematics* **116** (1982), pp. 113–132.

  A beautifully written, technical article on the existence of regular patterns on compact surfaces. Includes a very nice sketch of the history of these patterns and polyhedra.

- David W. Farmer and Theodore B. Stanford, *Knots and Surfaces: A Guide to Discovering Mathematics* (American Mathematical Society, 1996).

  An approach to many ideas presented in this book, based on completion of "tasks;" nice exercises.

- P. A. Firby and C. F. Gardiner, *Surface Topology* (Ellis Horwood Limited; 1982, 1991).

  Develops basic notions of surfaces and extends them to investigate vector fields, plane tessellations, and the fundamental group; includes discussion of plane Euclidean and hyperbolic geometry.

- Michael Henle, *A Combinatorial Introduction to Topology* (W. H. Freeman, 1979; Dover, 1994).

  A readable, gentle transition to more advanced topics, including vector fields, homology, covering spaces, and homotopy; some topics of point-set topology are also developed.

- T. Jensen and B. Toft, *Graph Coloring Problems* (John Wiley & Sons, 1995).

  Covers the coloring problems addressed in this book, in the context of coloring vertices of graphs, and many more such conjectures; a technical, advanced treatment with a tremendous bibliography.

- Charles Livingston, *Knot Theory* (Mathematical Association of America, 1993).

  An excellent presentation of a cross-section of topics in the theory of knots; prerequisites include basic linear algebra.

- William S. Massey, *Algebraic Topology: An Introduction* (Harcourt, Brace and World, 1967; Springer-Verlag, 1977).

  A standard, readable algebraic topology textbook suitable for advanced undergraduate students or graduate students; prerequisites include knowledge of some group theory.

- James R. Munkres, *Topology: A First Course* (Prentice-Hall, 1975).

  An excellent point-set topology textbook; includes some algebraic topology in the last chapter.

- V. V. Prasolov, *Intuitive Topology* (American Mathematical Society, 1995).

  An introduction to basic notions of topology through examples and problems, followed quickly by more sophisticated results, including vector fields, fixed point theorems, and periodic maps.

- Saul Stahl, "The other map coloring theorem," *Mathematics Magazine* **58** (1985), pp. 131–145.

  A wonderful survey of map coloring results and, in particular, the Heawood Conjecture.

- De Witt L. Sumners (editor), *New Scientific Applications of Geometry and Topology* (American Mathematical Society, 1992).

  Lecture notes from an A.M.S. short couse; includes quite readable papers which discuss scientific applications of topology, especially knot theory, in biology, chemistry, and physics.

- Jeffrey R. Weeks, *The Shape of Space* (Marcel Dekker, 1985).

  A wonderful and enlightening book which develops the notions of surface and 3-manifold through visualization exercises and entertaining discussion; good insightful introduction to geometries on surfaces and 3-manifolds.

- Magnus J. Wenninger, *Polyhedron Models* (Cambridge University Press, 1971).

  Describes how to construct models of all the regular and semi-regular polyhedral solids, in various forms; beautifully illustrated.

- A. White, *Graphs, Groups, and Surfaces* (North-Holland, 1973).

  A thorough treatment of many graph embedding problems and coloring conjectures; advanced, but includes many motivating illustrations.

- E. F. Whittlesey, "Finite surfaces: a study of finite 2-complexes," *Mathematics Magazine* **34** (1960), pp. 11–22, 67–80.

  An article that includes an extension of the classification theorem for compact surfaces based on circulation rules; quite readable and loaded with interesting hand-drawn illustrations.

- Robin J. Wilson, *Introduction to Graph Theory* (Academic Press, 1972).

  A good first book on graph theory; includes many worthwhile exercises.

Index